Ioanna Matilde Dimitriou

Symmetric Models, Singular Cardinal Patterns, and Indiscernibles

Ioanna Matilde Dimitriou

Symmetric Models, Singular Cardinal Patterns, and Indiscernibles

Infinitary combinatorics without the axiom of choice

Südwestdeutscher Verlag für Hochschulschriften

Impressum / Imprint
Bibliografische Information der Deutschen Nationalbibliothek: Die Deutsche Nationalbibliothek verzeichnet diese Publikation in der Deutschen Nationalbibliografie; detaillierte bibliografische Daten sind im Internet über http://dnb.d-nb.de abrufbar.

Bibliographic information published by the Deutsche Nationalbibliothek: The Deutsche Nationalbibliothek lists this publication in the Deutsche Nationalbibliografie; detailed bibliographic data are available in the Internet at http://dnb.d-nb.de.

Coverbild / Cover image: www.ingimage.com

Verlag / Publisher:
Südwestdeutscher Verlag für Hochschulschriften
ist ein Imprint der / is a trademark of
AV Akademikerverlag GmbH & Co. KG
Heinrich-Böcking-Str. 6-8, 66121 Saarbrücken, Deutschland / Germany
Email: info@svh-verlag.de

Herstellung: siehe letzte Seite /
Printed at: see last page
ISBN: 978-3-8381-3300-3

Zugl. / Approved by: Bonn, Univ., Diss., 2011

Dissertation angefertigt mit Genehmigung der Mathematisch-Naturwissenschaftlichen Fakultät Rheinischen Friedrich-Wilhelms-Universität Bonn

1. Gutachter: Prof. Dr. Peter Koepke

2. Gutachter: Prof. Dr. Benedikt Löwe

Tag der Promotion: 16. Dezember 2011

Summary

This thesis is on the topic of set theory and in particular large cardinal axioms, singular cardinal patterns, and model theoretic principles in models of set theory without the axiom of choice (ZF).

The first task is to establish a standardised setup for the technique of symmetric forcing, our main tool. This is handled in Chapter 1. Except just translating the method in terms of the forcing method we use, we expand the technique with new definitions for properties of its building blocks, that help us easily create symmetric models with a very nice property, i.e., models that satisfy the *approximation lemma*. Sets of ordinals in symmetric models with this property are included in some model of set theory with the axiom of choice (ZFC), a fact that enables us to partly use previous knowledge about models of ZFC in our proofs. After the methods are established, some examples are provided, of constructions whose ideas will be used later in the thesis.

The first main question of this thesis comes at Chapter 2 and it concerns patterns of singular cardinals in ZF, also in connection with large cardinal axioms. When we do assume the axiom of choice, every successor cardinal is regular and only certain limit cardinals are singular, such as $\aleph_\omega$.

Here we show how to construct almost any pattern of singular and regular cardinals in ZF. Since the partial orders that are used for the constructions of Chapter 1 cannot be used to construct successive singular cardinals, we start by presenting some partial orders that will help us achieve such combinations. The techniques used here are inspired from Moti Gitik's 1980 paper "All uncountable cardinals can be singular", a straightforward modification of which is in the last section of this chapter. That last section also tackles the question posed by Arthur Apter "Which cardinals can become simultaneously the first measurable and first regular uncountable cardinal?". Most of this last part is submitted for publication in a joint paper with Arthur Apter , Peter Koepke, and myself, entitled "The first measurable and first regular cardinal can simultaneously be $\aleph_{\rho+1}$, for any ρ". Throughout the chapter we show that several large cardinal axioms hold in the symmetric models we produce.

The second main question of this thesis is in Chapter 3 and it concerns the consistency strength of model theoretic principles for cardinals in models of ZF, in connection with large cardinal axioms in models of ZFC. The model theoretic principles we study are variations of Chang conjectures, which, when looked at in models of set theory with choice, have very large consistency strength or are even inconsistent.

We found that by removing the axiom of choice their consistency strength is weakened, so they become easier to study. Inspired by the proof of the equiconsistency of the existence of the ω_1-Erdős cardinal with the original Chang conjecture, we prove equiconsistencies for some variants of Chang conjectures in models of ZF with various forms of Erdős cardinals in models of ZFC. Such equiconsistency results are achieved on the one direction with symmetric forcing techniques found in Chapter 1, and on the converse direction with careful applications of theorems from core model theory. For this reason, this chapter also contains a section where the most useful 'black boxes' concerning the Dodd-Jensen core model are collected.

More detailed summaries of the contents of this thesis can be found in the beginnings of Chapters 1, 2, and 3, and in the conclusions, Chapter 4.

Acknowledgements

Doctoral theses take a long time to prepare, and this process is not as smooth as one might hope or plan. With this note I would like to thank the people who played a crucial role in this process.

I would like to thank first and foremost my parents Christos and Marta, for their loving support, also financial but mainly emotional, and for the safety net that they provide me with. An essential role in my emotional stability played also my partner Calle Runge, whom I thank from the bottom of my heart for his emotional support, for providing me with a homely environment to escape to, and for our amazing conversations about life, the universe, and everything. Also when stress got way too much, he was there to help me, along with Jonah Lamp and Mark Jackson, who with me are "Jonah Gold and his Silver Apples". Guys, thank you so much for all the fun and stress relief that drumming in a band gave me in the difficult times of a PhD.

I am grateful to my supervisor Peter Koepke for the opportunity to fulfill this dream of mine, and for all that he taught me, not only in the realm of core model theory, but also in the social aspects of the academic world. A big thank you goes also to Benedikt Löwe, the second referee of this thesis, for his invaluable advice ever since my master's studies at the ILLC in Amsterdam, and for his question on patterns of singular cardinals that led to Section 2 of Chapter 2. Special thanks go to Andrés Caicedo and to my officemate Thilo Weinert, for their suggestions and corrections of earlier versions of the thesis, and a big thanks to the Mathematical logic group of Bonn for listening to my detailed presentations of Chapters 1 and 2. I also thank Arthur Apter for coauthoring a paper with Peter Koepke and myself, and for the informative discussions. I am grateful to all the people in the "Young set theory workshop" for the engaging talks, formal and informal.

I would like to express my appreciation to Moti Gitik whose ideas, especially in his paper "All uncountable cardinals can be singular" have kept me busy for a good two years during my PhD studies, and have taught me so much. Also a special thanks for a short comment he gave me during the workshop "Infinitary combinatorics without the axiom of choice" which led me to really understand why there is no easier way to achieve his result.

I thank my sister Yolanda for the bond we have, and my friends in Bonn for changing my view of this city, and making it a home for me. And lastly, I want to thank Malaclypse the younger for writing Principia Discordia, the book that, apart from just being awesome, kept reminding me that life is not as gloomy and difficult as it some times seems.

Contents

–There are trivial truths and there are great truths. The opposite of a trivial truth is plainly false. The opposite of a great truth is also true.

Niels Bohr

0 Introduction

This thesis is on set theory and it is part of the investigation of large cardinal axioms and of cardinal patterns in the absence of the axiom of choice. In particular, it is a study of singular cardinal patterns and of infinitary combinatorial and model theoretic principles under the light of set theoretic techniques such as symmetric forcing and theorems from core model theory.

In an attempt to explain this topic and to give the motivation behind this project let us take a look at some history on the subject, to see how these techniques and questions came to be, and how do they connect to the questions posed in this thesis. This initial part of the introduction does not require any prior knowledge of set theory. The next section will.

The story of axiomatic set theory starts with Georg Cantor's proof that there are strictly more real numbers than natural numbers, the potential contradictions in his formulation of set theory and the quest to find the "true" axioms that will rid mathematics of these contradictions, thus making it a legitimate candidate for the foundations of mathematics. The biggest problem in Frege's formalisation of Cantor's intuition of this early set theory was that full comprehension was used to make sets, i.e., for any formula (or property) φ one could form the sets of all things with property φ.

The problem with this is illustrated best in a paradox discovered simultaneously by Bertrand Russell and Ernst Zermelo, the famous "Russell's paradox". The paradox arises when one defines the set A of all things that are not elements of themselves, and asks the question "Does A belong to A?". This is the same as the well known "barber paradox", found in [**Rus19**]. There Russell writes the following.

> You can define the barber as "one who shaves all those, and those only, who do not shave themselves". The question is, does the barber shave himself? In this form the contradiction is not very difficult to solve. But in our previous form [the original Russell's paradox] I think it's clear that you can only get around it by observing that the whole question whether a class is or is not a member of itself is nonsense, i.e., that no class is or is not a member of itself, and that it is not even true to say that, because the whole form of words is just noise without meaning.

What I am trying to do is underline that paradoxes were a big deal for the mathematicians of the time. Set theorists wanted this to be the foundation of all mathematics, and in such a foundation there is no room for contradictions! In response to all this Ernst Zermelo, among others, attempted an axiomatic setting for set theory. In this attempt [**Zer08**] he formulated the axiom of choice (AC). This axiom can be stated as follows.

> For every collection $\mathcal{A}$ of pairwise disjoint non-empty sets, there is a set that contains exactly one element from each of the sets in $\mathcal{A}$.

Even after the axiomatisation of modern set theory was completed, with additional axioms by Abraham Fraenkel and Alfred Tarski [**Fra22b**], to form the axiomatic system of Zermelo-Fraenkel set theory ZFC, the axiom of choice remained controversial as it implies many counterintuitive theorems. A famous such theorem is the Banach-Tarski paradox, which uses the axiom of choice to split a solid ball into finitely many disjoint parts, which can be then rearranged to form two solid balls, identical to the original one. Note that the reassembly is done only with moving and rotating these pieces, not by stretching them in any way. As we read in the Stanford Encyclopedia of Philosophy [**Jec02**, §5],

> This is of course a paradox only when we insist on visualizing abstract sets as something that exists in the physical world. The sets used in the Banach-Tarski paradox are not physical objects, even though they do exist in the sense that their existence is proved from the axioms of mathematics (including the Axiom of Choice).

Equivalences between AC and other perhaps counterintuitive mathematical theorems such as the wellordering principle and Zorn's lemma fuelled the debate on the validity of the axiom of choice. But one may joke[1]:

> The Axiom of Choice is obviously true, the wellordering principle obviously false, and who can say about Zorn's lemma?

In all the controversy of the early days of set theory, Fraenkel in 1922 [**Fra22a**] proved the independence of AC from a slightly weaker version of set theory, that is set theory with atoms (ZFA). Atoms are sets with no elements in them but different from the empty set. To prove this result he used the permutation method which was refined by Adolf Lindenbaum and

[1]This folklore joke is often attributed to American mathematician Jerry L. Bona

Andrzej Mostowski [**LM38**] in order to prove the independence of weaker forms of AC from set theory with atoms.

The consistency of AC relative to ZF itself was shown by Kurt Gödel in 1938 [**Göd38**] by constructing a model of set theory with the axiom of choice (ZFC) by starting from a model of set theory without the axiom of choice (ZF). In a sense, a theory being consistent is the same as the theory having a model (a collection of sets that satisfy all the axioms of the theory). For this proof, Gödel created the model of all constructible sets L. Gödel's construction evolved in the 70s to what is now known as core model theory and this is still a hot topic in modern set theoretic research. This thesis contains only black-box style usage of theorems for an early core model, developed mostly in the 70s and 80s, the Dodd-Jensen core model.

The consistency of the negation of the axiom of choice ($\neg$AC) and therefore its independence from the the other axioms of ZF, was proved in 1963 by Paul Cohen [**Coh63**] with his newly invented method of forcing. In that paper he used arguments from Fraenkel's permutation method. With the work of Solovay [**Sol63**], Dana Scott [**Sco67**], Vopěnka (e.g., [**Vop65**]), Vopěnka and Hájek [**VH67**], and Thomas Jech [**Jec71**], these arguments were turned into the symmetric forcing technique in terms of forcing with Boolean valued models. Nowadays, forcing is more often carried out with partial orders instead of Boolean valued models. In the next chapter there is an exposition of this symmetric forcing technique, in terms of partial orders, in a very general setting and with theorems and definitions that make it quite simple to construct models of ZF + $\neg$AC (set theory with the negation of the axiom of choice), called symmetric models, which have very nice properties such as the "approximation lemma". When a symmetric model satisfies the approximation lemma, its sets of ordinals are also in some smaller generic extension that satisfies the axiom of choice.

In 1931, even before the independence of AC from ZF was shown, and before his construction of L, Gödel proved his famous incompleteness theorems (see [**Göd30**] and [**Göd31**]) which changed set theory and logic for ever. In particular, Gödel's second incompleteness theorem is the following fact.

> Any consistent axiomatic theory that can interpret elementary arithmetic cannot prove its own consistency.

In other words, if a "reasonable" theory proves that it itself is consistent, then it must be inconsistent! This had a huge impact at the time and it convinced set theorists that they can give up trying to prove that ZF is consistent. So the only thing left to do in that direction is to get evidence of consistency by studying stronger and stronger axioms.

What later was the first such strong axiom was first defined by Bonner mathematician Felix Hausdorff during his time in Leipzig in 1908[2], and it is known as the existence of a

[2] In fact, this paper was at the same volume of Mathematische Annalen that Zermelo's first axiomatisation of set theory was published!

weakly inaccessible cardinal [**Hau08**]. In his famous "Grundzüge der Mengenlehre" [**Hau14**] Hausdorff commented on the largeness of such a cardinal by saying:

> ... ist die kleinste unter ihnen von einer so exorbitanten Größe, daß sie für die üblichen Zwecke der Mengenlehre kaum jemals in Betracht kommen wird.

Of course he was wrong in that last prediction, since already by 1939 these, and even larger ones, were indeed used. For example, Gödel in [**Göd39**, Theorem 8] worked with inaccessibles to prove that if there exists a weakly inaccessible cardinal then there exists a set that is a model of set theory which has no weakly inaccessible cardinals in it.

Large cardinals appear often in what is called *relative consistency strength* proofs. For these proofs one starts from the assumption that ZFC+"axiom X" is consistent, so it has a model, and from that model one constructs another that satisfies ZFC+"axiom Y". Such an argument proves that "axiom X" is of greater or equal consistency strength than "axiom Y", relative to the axioms of ZFC. Axioms that are usually considered in place of "axiom X" or "axiom Y", are often statements asserting the existence of cardinals with certain properties (e.g., being a weakly inaccessible cardinal). These are called 'large cardinal properties', and the axioms 'large cardinal axioms'. To this day, this is a very rich field in set theory with many relative consistency strength results, especially for set theory *with* the axiom of choice.

This thesis has relative consistency strength results of the form

If the system ZFC+"axiom X" is consistent

then the system $\mathsf{ZF} + \neg\mathsf{AC}$+"axiom Y" is consistent.

And results of the opposite direction. There are several reasons for this. Apart from being intrinsically interesting, these constructions help weaken the consistency strength of some large cardinal axioms, so that they can be studied with existing techniques. Some of these axioms are too strong or even inconsistent with the axiom of choice, but when one looks at them in a choiceless environment they become weaker and easier to handle.

There is an axiom of a different nature to large cardinal axioms, that contradicts the axiom of choice, yet it is quite popular among modern set theorists; that is the axiom of determinacy (AD). This is a statement about games, and the existence of winning strategies for certain two-player games. The axiom of determinacy contradicts the axiom of choice and it implies that sets of real numbers have "nice" properties in some sense. The system of ZF with AD is much stronger than the system ZFC in terms of consistency strength. So consistency strength studies between systems of the form ZFC+"axiom X" and ZF+"axiom Y", help determine the consistency strength of patterns of cardinals with combinatorial properties that occur naturally in models of $\mathsf{ZF} + \mathsf{AD}$, but are impossible in models of ZFC.

In this thesis, to carry out such proofs, in the direction from having a model of ZFC+"axiom X" to getting a model of ZF+"axiom Y", I used the aforementioned technique of symmetric forcing. By working with this technique for the past six years (starting during my master's thesis in the Institute of Logic, Language, and Computation in Amsterdam), I have somewhat standardised it, in order to make it easily implementable in an array of results, and make things easier for myself. As you'll hopefully read later in the thesis, this lead to several models constructed with different combinations of just two forcing techniques and two sorts of "symmetry generators", in various "lengths".

In Chapter 2 we include ideas from Moti Gitik's 1980 paper "All uncountable cardinals can be singular" [**Git80**], a paper with which I spent some years during my PhD project, a paper whose complex ideas really impressed me in the past years. I would love to see these ideas implemented to even more complex forcing constructions than the ones presented here.

On the other hand, most symmetric forcing constructions in Chapter 3 are not so impressive (any more), they are mostly simple implementations of a very early symmetric model. Underlining this chapter are the techniques of core model theory, here used in a black-box fashion.

From Ronald Jensen, who together with Tony Dodd created the first "core model" that is stronger than L (i.e., it may satisfy more large cardinal axioms than L can), to my Doktorvater Peter Koepke, his PhD student Ralf Schindler, and others, core model theory and its applications to relative consistency strength analysis have a long history in the mathematical logic group of Bonn. This thesis does not directly add to core model theory itself but uses a known clever manoeuvre to define the core model inside a model of ZFC, even if we only have a model of ZF. This method is used in order to take what properties we can from our 'choiceless' model and pass them to the core model which will be inside a model of ZFC. Then, usually by following to some extent proofs from the literature for models of ZFC with these properties, we produce a model that satisfies certain large cardinal axioms.

Recapping, this is how we achieve an equiconsistency result: From a model of $\mathsf{ZFC}+$"axiom X" we use symmetric forcing to get a model of $\mathsf{ZF}+\neg\mathsf{AC}+$"axiom Y". Then, we start from a model of $\mathsf{ZF}+\neg\mathsf{AC}+$"axiom Y", pass some of its properties to a core model inside it, and use the core model's structured nature to reconstruct some of the strength of "axiom X", thus getting a model of $\mathsf{ZFC}+$"axiom X". We say then that the theories $\mathsf{ZFC}+$"axiom X" and $\mathsf{ZF}+\neg\mathsf{AC}+$"axiom Y" are equiconsistent.

Before we go to the preliminaries, I would like to end this historical note in a more personal note. It is my view of science and all of all human endeavours that, apart from the set theoretic universe V, nothing is made out of nothing, and that all the ideas are evolutions of previous ideas. As an example, the idea of infinity was around since the time of the ancient Greeks, as we know from the Aristotle's records of Zeno of Elea's paradoxes (450 BC). Also the ancient Indian mathematical text *Surya Prajnapti* (ca.400 BC) includes an impressive classification of finite and infinite sets, the later subdivided into "nearly infinite, truly infinite, infinitely infinite". Medieval mathematicians such as Albert of Saxony proves theorems about infinite sets, and by the 1847, Bolzano was a firm supporter of the usefulness of infinite sets in Mathematics. Bolzano in fact proved that every infinite set can be put in a 1-1 correspondence with one of its proper subsets [**Bol51**]. I say this not to give an exposition of the long history of the idea of infinity, but just to give some indication as to why I believe that no idea is entirely of one person. For this reason, and also in order to include the reader in this investigation, this thesis is from now on entirely written on first person plural.

1. Zermelo Fraenkel set theory

From now on we assume that the reader has some basic knowledge of first order logic and set theory, as laid for example in [**Jec03**, Part I], [**Kun80**, Chapter I], and [**Hod97**]. We will define the notions we will mostly use in this thesis in order to fix our notation and to make the thesis as self contained as possible.

This thesis is written under the assumption of ZFC, therefore we list its axioms:

Extensionality	If x and y have the same elements then they are equal.
Foundation	Every non-empty set has an $\in$-minimal element.
Comprehension schema	For every formula φ with free variables among x, z, w, if z and w are sets then $\{x \in z\ ;\ \varphi(z, x, w)\}$ is a set.
Pairing	If x and y are sets, then $\{x, y\}$ is a set.
Union	For every set x there exists the set $y = \bigcup x$, the union of all elements of x.
Replacement schema	For every formula φ with z, y amongst its free variables, if φ is function-like, and x is a set, then $\{y\ ;\ \exists z \in x(\varphi(z, y))\}$ is a set.
Infinity	There exists an infinite set.
Powerset	For every set x there exists a set $y = \mathcal{P}(x)$, the set of all subsets of x.
Choice	Every set of non-empty sets has a choice function.

Let us start with a quick word on the notation we will be using, in order to avoid long definitions of basic notions. By $\mathsf{ot}(x)$ we mean the ordertype of x. If α, β are ordinals, then (α, β) is the open ordinal interval α, β, i.e., all the ordinals greater than α and smaller than β. We often view sequences as functions with domain a subset of the ordinals. We write $\mathsf{dom}(f)$ for the domain of a function or a sequence f and $\mathsf{rng}(f)$ for its range. A partial function is denoted by $\rightharpoonup$, i.e., $f : x \rightharpoonup y$ is a function from a subset of x to y. For the minimal element of a pair of ordinals we may write $\mathsf{min}(\alpha, \beta)$, but we may also write $\mathsf{min} X$ for a set of ordinals X and mean the least element of X. We denote the class of all ordinals by Ord.

The axiom of choice (AC) and the notion of cardinality are of particular interest in this thesis. For the purposes of this thesis, when we are working in ZF and we say "the cardinality of a set x", or equivalently write $|x|$, we imply that the set x is wellorderable. Therefore we consider only initial ordinals as cardinals in this thesis. Now let us take a closer look at the axiom of choice. If x is a set of non-empty sets, then a choice function for x is a function f such that for every $y \in x$, $f(y) \in y$, i.e., f picks an element from every $y \in x$. It is well known that the following are equivalent (see e.g., [**Jec73**, Theorem 2.1]):

- The axiom of choice,
- the wellordering principle, i.e., the statement "every set can be well-ordered", and
- Zorn's lemma, i.e., the statement "for every non-empty partial order $\langle \mathbb{P}, \leq \rangle$, if every chain in $\mathbb{P}$ has an upper bound, then $\mathbb{P}$ has a maximal antichain".

Since we will be constructing models of $\mathsf{ZF} + \neg\mathsf{AC}$ it will be interesting to know how much choice does hold in those models, or how much choice fails. We will now look at some of these fragments of choice and what they imply. A weaker form of the axiom of choice that we will be looking at is the following.

Definition 0.1. For sets A and B, $\mathsf{AC}_A(B)$ is the statement "for every set X of nonempty subsets of B, if there is an injection from X to A then there is a choice function for X.

For example, $\mathsf{AC}_\omega(\mathbb{R})$ translates to "every countable set of non-empty subsets of the reals has a choice function". An important consequence of this choice fragment is the following.

Lemma 0.2. ($\mathsf{AC}_A(B)$) *If there is a surjection from B to A then there is an injection from A to B.*

Proof. Let A, B be arbitrary non-empty sets, and let $f : B \to A$ be a surjection. For every $a \in A$ let $X_a \overset{\text{def}}{=} \{b \in B \,;\, f(b) = a\}$. Let g be a choice function of $X \overset{\text{def}}{=} \{X_a \,;\, a \in A\}$. Define $h : A \to B$ by $h(a) = g(X_a)$. This is an injection because f is a function. qed

The converse is not true, i.e., if a surjection from B to A implies an injection from A to B then it is not necessarily the case that $\mathsf{AC}_A(B)$ holds. For example, there are models of ZF where $\mathsf{AC}_\omega(\mathbb{R})$ fails (e.g., the Feferman-Levy model) yet it is a theorem of ZF that there is a surjection from $\mathbb{R}$ to ω and an injection from ω to $\mathbb{R}$.

In our choiceless constructions we will always end up with some non-wellorderable powerset of a cardinal. So we will be looking at fragments of choice of the form $\mathsf{AC}_{\kappa^+}(\mathcal{P}(\kappa))$ and $\mathsf{AC}_\kappa(\mathcal{P}(\kappa))$, for an infinite wellordered cardinal κ. An easy fact is the following.

Lemma 0.3. (ZF) *For every infinite cardinal κ, there is a surjection from $\mathcal{P}(\kappa)$ onto κ^+.*

For a proof see for example [**Dim06**, Lemma 1.4]. In every symmetric model we will see in this thesis, the powerset of a certain cardinal κ will be either a κ^+-long union of $\leq \kappa$-sized sets or even worse, a κ-long union of κ-sized sets. Let's define the following 'anti-choice' fragment.

$$\mathsf{SUS}(\kappa) \overset{\text{def}}{\iff} \text{"The powerset of } \kappa \text{ is a } \kappa\text{-sized union of sets of cardinality } \leq \kappa\text{"},$$

where SUS stands for "Small Union of Small sets". According to [**Dim06**, Lemma 2.6], if a powerset of a cardinal is such a small union of small sets then that powerset is small in a surjective way.

LEMMA 0.4. (ZF) *The statement* $\mathsf{SUS}(\kappa)$ *implies that there is no surjection from* $\mathcal{P}(\kappa)$ *onto* $(\kappa^+)^\kappa$.

Using Lemma 0.3 and Lemma 0.4 it's easy to show the following (see e.g., [**Dim06**, Theorem 2.7]).

LEMMA 0.5. (ZF) *The statement* $\mathsf{SUS}(\kappa)$ *implies that* κ^+ *is singular and that every wellorderable subset of* $\mathcal{P}(\kappa)$ *has size at most* κ.

This means that if $\mathsf{SUS}(\kappa)$ holds then the powerset of κ is small in an injective way as well. Similarly to [**HKR**$^+$**01**, Theorem 2] we can show the following.

LEMMA 0.6. (ZF) *Under* $\mathsf{AC}_\kappa(\mathcal{P}(\kappa))$, κ^+ *is regular.*

As a corollary, $\mathsf{AC}_\kappa(\mathcal{P}(\kappa))$ implies $\neg\mathsf{SUS}(\kappa)$. We have the following implications between these choice forms.

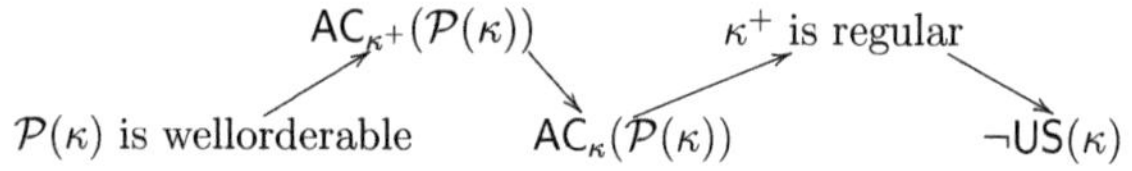

So, if in a model of ZF we prove the negation of any of the last four statements then we would have proved that the powerset of κ is non-wellorderable.

2. Large cardinals

Throughout this thesis we will be looking at large cardinals under the negation of AC. In terms of consistency strength, the first large cardinal axiom we consider is the existence of inaccessible cardinals. Recall that an inaccessible cardinal is a regular cardinal κ such that for every $\alpha < \kappa$, $2^\alpha < \kappa$. This is often called strongly inaccessible in the literature, with weakly inaccessible being just a regular limit cardinal with no requirements on the size of any powersets. Here we will always say inaccessible cardinal and mean strongly inaccessible cardinal. The topic of inaccessible cardinals with the negation of AC was considered in [**Dim06**] and [**BDL06**]. Here we will only assume inaccessible cardinals in the context of ZFC. The first large cardinal axiom that we will encounter under $\neg$AC is the existence of measurable cardinals.

In ZFC, there are many equivalent ways to define a measurable cardinal (see e.g. [**Kan03**, §2]). We define a measurable cardinal as follows.

DEFINITION 0.7. A cardinal κ is measurable iff there exists a non-trivial, κ-complete uniform ultrafilter over κ.

In ZFC this is equivalent to saying that κ has a uniform normal measure, i.e., an ultrafilter U such that for any sequence $\langle X_\alpha \,;\, \alpha < \kappa\rangle \in {}^\kappa U$, its diagonal intersection

$$\triangle_{\alpha<\kappa} X_\alpha \stackrel{\text{def}}{=} \{\xi < \kappa \,;\, \xi \in \bigcap_{\alpha<\xi} X_\alpha\}$$

is in U. In ZF we have the following.

LEMMA 0.8. (ZF) *An ultrafilter U over a cardinal κ is normal iff for every regressive $f : \kappa \to \kappa$, there is an $X \in U$ such that f is constant on X.*

PROOF. "$\Rightarrow$" Assume for a contradiction that for every $\alpha < \kappa$, $f^{-1}``\{\alpha\} \notin U$. Since $U$ is an ultrafilter, this means that for every $\alpha < \kappa$, $\kappa \setminus f^{-1}``\{\alpha\} \in U$. By the normality of U, the diagonal intersection $\triangle_{\alpha<\kappa}(\kappa \setminus f^{-1}``\{\alpha\})$ is in $U$ and therefore it is not empty. But then there is some $\xi \in \triangle_{\alpha<\kappa} X_\alpha$, i.e., $\xi \in \bigcup_{\alpha<\xi}(\kappa \setminus f^{-1}``\{\alpha\})$, which means that for every $\alpha < \xi$, $f(\xi) \neq \alpha$, so $f(\xi) \geq \xi$, which is impossible since f is regressive.

"$\Leftarrow$" Let $\langle X_\alpha \,;\, \alpha < \kappa\rangle \in {}^\kappa U$ and assume for a contradiction that $\triangle_{\alpha<\kappa} X_\alpha \notin U$, so $\kappa \setminus \triangle_{\alpha<\kappa} X_\alpha \in U$. Define a function $f : \kappa \to \kappa$ by $f(\xi) \stackrel{\text{def}}{=} \min\{\alpha < \kappa \,;\, \xi \notin X_\alpha\}$ if $\xi \notin \triangle_{\alpha<\kappa} X_\alpha$ and $f(\xi) = 0$ otherwise. The function f is regressive so there is an $X \in U$ such that f is constant in X. So there is an $\alpha < \kappa$ such that $f^{-1}``\{\alpha\} \in U$. Since $X_\alpha \in U$, $f^{-1}``\{\alpha\} \cap X_\alpha \in U$. Thus for some $\beta < \kappa$, $\beta \in f^{-1}``\{\alpha\} \cap X_\alpha$. But $\beta \in f^{-1}``\{\alpha\}$ implies that $\beta \notin X_\alpha$ which is impossible. qed

Jech [**Jec68**] and Takeuti [**Tak70**] independently showed that if we assume the consistency of ZFC+"there exists a measurable cardinal" then the theory ZF+"ω_1 is a measurable cardinal" is consistent. In Section 1.33 we will give Jech's construction modified so that the construction can be carried out in generality to give a model where the measurable cardinal becomes any predetermined successor of a regular cardinal while staying a measurable cardinal. The generality of this construction and the lemmas around it help us combine it with other structures or use it with other large cardinal axioms in order to give a wide range of results.

In the topic of measurable cardinals we will be also looking at the question posed by Arthur Apter "Which cardinals can become simultaneously the first measurable and first regular cardinal?", in combination with the author's modification of Gitik's model in [**Git80**]. That will be studied in the last section of Chapter 2, it is joint work with Peter Koepke and Arthur Apter, and it is submitted for publication [**ADK**]. That construction requires strongly compact cardinals.

DEFINITION 0.9. If κ is a cardinal and $\alpha \geq \kappa$ an ordinal, we say that a filter F over $\mathcal{P}_\kappa(\alpha) \stackrel{\text{def}}{=} \{x \subseteq \alpha \,;\, |x| < \kappa\}$ is fine iff F is κ-complete, and for any $\xi \in \alpha$, $\{x \in \mathcal{P}_\kappa(\alpha) \,;\, \xi \in x\} \in F$.

A cardinal κ is called α-strongly compact iff there exists a fine ultrafilter over $\mathcal{P}_\kappa(\alpha)$. A cardinal κ is called strongly compact iff it is α-strongly compact for every $\alpha \geq \kappa$.

According to [**Kan03**, Corollary 4.2], strongly compact cardinals are measurable cardinals, therefore limits of Ramsey cardinals and limits of inaccessible cardinals. We will use this fact in Chapter 2 when we want to collapse certain intervals between strongly compact cardinals. The models in that chapter will be built from strongly compact cardinals. After the construction strong compactness will be destroyed but some combinatorial properties will remain. We will have a lot of almost Ramsey cardinals around and some Rowbottom cardinals. To define such large cardinals we need to define the following partition relations.

DEFINITION 0.10. For ordinals $\alpha, \beta, \gamma, \delta$, the partition relation

$$\beta \to (\alpha)^{\gamma}_{\delta}$$

means that for any $f : [\beta]^{\gamma} \to \delta$ there is an $X \in [\beta]^{\alpha} \overset{\text{def}}{=} \{x \subseteq \beta \,;\, \mathsf{ot}(x) = \alpha\}$ such that X is homogeneous for f, i.e., $|f``[X]^{\gamma}| = 1$.

For ordinals α, β, γ, the partition relation

$$\beta \to (\alpha)^{<\omega}_{\gamma}$$

means that for any $f : [\beta]^{<\omega} \to \gamma$, where $[\beta]^{<\omega} \overset{\text{def}}{=} \bigcup_{n\in\omega}[\beta]^n$, there exists an $X \in [\beta]^{\alpha}$ that is homogeneous for $f{\restriction}[\beta]^n$, i.e., for every $n \in \omega$, $|f``[X]^n| = 1$.

The first partition relation immediately gives rise to the following definition.

DEFINITION 0.11. A cardinal $\kappa > \omega$ is weakly compact iff $\kappa \to (\kappa)^2_2$.

We will look only briefly at weakly compact cardinals in connection with Jech's model where a successor cardinal is measurable. A useful observation in connection also with measurable cardinals is the following.

LEMMA 0.12. (ZF) *If κ is measurable with a normal measure or κ is weakly compact and $\alpha < \kappa$, then there is no injection $f : \kappa \to \mathcal{P}(\alpha)$.*

This is [**Bul78**, Proposition 0.2]. The following is [**Bul78**, Corollary 0.3].

COROLLARY 0.13. (ZF) *If κ is a successor cardinal then AC_κ implies that κ is neither measurable with a normal measure nor it is weakly compact.*

In [**Kan03**, Theorem 7.8] we see that weakly compact cardinals are inaccessible in ZFC. As we see in that theorem, weakly compact cardinals have many definitions which are equivalent in ZFC. Although it would be interesting to separate some of them in ZF, this is not in the scope of this thesis. The following large cardinals notions will play a more important role in for us here.

DEFINITION 0.14. For an infinite ordinal α, the (α-)Erdős cardinal $\kappa(\alpha)$ is the least κ such that $\kappa \to (\alpha)^{<\omega}_2$.

When we say just "Erdős cardinal" we refer to a cardinal that is the α-Erdős cardinal $\kappa(\alpha)$, for some ordinal α. We will look at Erdős cardinals in detail in Chapter 3 in connection with structures and elementary substructures (see the next section). We will heavily use the generalised Jech model with them. For that we will use the following standard fact a lot.

LEMMA 0.15. (ZFC) *Suppose that $\alpha \geq \omega$ is a limit ordinal and assume the Erdős cardinal $\kappa(\alpha)$ exists. Then*

(a) *For any $\gamma < \kappa(\alpha)$, $\kappa(\alpha) \to (\alpha)^{<\omega}_{\gamma}$.*
(b) *$\kappa(\alpha)$ is inaccessible.*

This is [**Kan03**, Proposition 7.14]. The fixed points of the sequence of Erdős cardinals are called Ramsey cardinals.

DEFINITION 0.16. A cardinal κ is a Ramsey cardinal iff $\kappa \to (\kappa)^{<\omega}_2$.

Ramsey cardinals are inaccessible in ZFC because they are Erdős. Measurable cardinals are Ramsey in ZFC [**Kan03**, Corollary 7.18] and in fact if U is a normal measure for a cardinal κ, then the set $\{\alpha < \kappa \, ; \, \alpha \text{ is Ramsey}\} \in U$ [**Kan03**, Exercise 7.19]. Just as with Erdős cardinals, we have the following.

LEMMA 0.17. (ZFC) *A cardinal κ is Ramsey iff for any $\gamma < \kappa$, $\kappa \to (\kappa)^{<\omega}_{\gamma}$.*

In the models of Chapter 2 we will use Ramsey cardinals in inner models to get cardinals with the following property that is a weakening of $\kappa \to (\kappa)^{<\omega}_2$.

DEFINITION 0.18. If κ is a cardinal and for every $\alpha < \kappa$ $\kappa \to (\alpha)^{<\omega}_2$ holds then κ is called an almost Ramsey cardinal.

In ZFC almost Ramsey cardinals are strong limit cardinals [**AK08**, Proposition 2]. Apter and Koepke worked on almost Ramsey cardinals in choiceless models in [**AK08**] where they studied the consistency strength of the existence of a class of almost Ramsey cardinals.

Two weaker versions of the partition relation $\beta \to (\alpha)^{\gamma}_{\delta}$ are the following.

DEFINITION 0.19. For ordinals $\alpha, \beta, \gamma, \delta$, the square brackets partition relation

$$\beta \to [\alpha]^{<\omega}_{\gamma}$$

means that for every $f : [\beta]^{<\omega} \to \gamma$, there is an $X \in [\beta]^{\alpha}$ such that $f``[X]^{<\omega} \neq \gamma$. The square brackets partition relation

$$\beta \to [\alpha]^{<\omega}_{\gamma,<\delta}$$

means that for every $f : [\beta]^{<\omega} \to \gamma$, there is an $X \in [\beta]^{\alpha}$ such that $|f``[X]^{<\omega}| < \delta$.

With this we can now define Rowbottom cardinals.

DEFINITION 0.20. A cardinal κ is a Rowbottom cardinal iff for any $\gamma < \kappa$, $\kappa \to [\kappa]^{<\omega}_{\gamma,<\omega_1}$.

By Lemma 0.17 we have that in ZFC every Ramsey cardinal is Rowbottom.

DEFINITION 0.21. A filter F over a cardinal κ is a Rowbottom filter iff for every $\gamma < \kappa$ and every $f : [\kappa]^{<\omega} \to \gamma$, there is an $X \in F$ such that $|f``[X]^{<\omega}| < \omega_1$. If such an F exists then κ is a Rowbottom cardinal carrying a Rowbottom filter.

We will encounter such cardinals in Chapter 2. Arthur Apter has posed the question "How large is the least Rowbottom cardinal?" [**Apt83b**]. Silver conjectured that, even assuming AC, it is consistent that $\aleph_\omega$ is the least Rowbottom cardinal but that remains unproven to this day. Starting from a model of ZFC+"there exists countable sequence of measurable cardinals" Apter constructed a model of ZF + DC+"$\aleph_\omega$ is a Rowbottom cardinal carrying a Rowbottom filter" [**Apt83b**, Theorem 1]. Later in a joint paper of Apter with Peter Koepke [**AK06**] this result was improved and an equiconsistency was achieved. They proved that for every natural numbers n the theory $\mathsf{ZF}+\mathsf{DC}_{\aleph_n}+$"$\aleph_\omega$ is a Rowbottom cardinal carrying a Rowbottom filter" is equiconsistent with the theory ZFC+"there exists a measurable cardinal". With this methods we were able to prove that in the constructions in Chapter 2 that start from models of ZFC with strongly compact cardinals, some symmetric models have many Rowbottom cardinals carrying Rowbottom filters. These are only the limit cardinals in a certain interval.

A notion similar to the notion of a Rowbottom cardinal is the following.

DEFINITION 0.22. A cardinal κ is called Jónsson iff $\kappa \to [\kappa]^{<\omega}_\kappa$.

In the same paper of Apter and Koepke where $\aleph_\omega$ and $\aleph_{\omega_1}$ are made Rowbottom, we see that the existence of certain countable sequences of Erdős cardinals with coherence properties imply that their supremum is a Jónsson cardinal. That Jónsson cardinal would have cofinality ω, which by [**AK06**, Theorem 6] means there is then an inner model with a measurable cardinal. By [**AK06**, Theorem 3] those sequences of Erdős cardinals can be forced by starting from a measurable cardinal, therefore Jónsson cardinals will be our stepping stone towards this equiconsistency result.

This study of variants of Erdős cardinals in Chapter 3 has to do with Chang conjectures (which we will define and look at in that chapter) and properties of first order structures. For example, one of the equivalent definitions of Jónsson cardinals is the statement "every first order structure of size κ with a countable language has a proper elementary substructure of the same cardinality (for the equivalence see [**Kan03**, Exercise 8.12]). In fact, this statement is implied by $\kappa \to [\kappa]^{<\omega}_2$ in ZF. We will see a lot of arguments of this sort in Chapter 3 but for that we need to have some definitions from model theory.

3. Structures and elementary substructures

In Chapter 3 we will be looking at first order structures with wellorderable domains, i.e., of the form $\mathcal{A} = \langle A, \dots \rangle$ for some wellorderable A, and with countable languages. Let $\mathcal{A} = \langle A, \dots \rangle$ and $\mathcal{B} = \langle B, \dots \rangle$ be wellorderable structures with the same language $\mathcal{L}$. We say that a function $j : A \to B$ is an elementary embedding of $\mathcal{A}$ to $\mathcal{B}$, and we write $j : \mathcal{A} \prec \mathcal{B}$, iff for any formula $\varphi(v_1, \dots, v_n)$ of $\mathcal{L}$ and for any $x_1, \dots, x_n \in A$,

$$\mathcal{A} \models \varphi(x_1, \dots, x_n) \iff \mathcal{B} \models \varphi(j(x_1), \dots, j(x_n)).$$

If the above holds only for Σ_1-formulas then we say that j is a Σ_1-elementary embedding, and we write $j : \mathcal{A} \prec_1 \mathcal{B}$.

If $A \subseteq B$ and the identity is an elementary embedding then we say that $\mathcal{A}$ is an elementary substructure of $\mathcal{B}$ and we write $\mathcal{A} \prec \mathcal{B}$.

The easiest way to get an elementary substructure is by taking Skolem hulls. For a wellorderable first order structure $\mathcal{A} = \langle A, \ldots \rangle$ with a countable language $\mathcal{L}$, and for a formula $\varphi(y, v_1, \ldots, v_n)$ of $\mathcal{L}$ we say that a function $f : A^n \to A$ is a Skolem function for φ iff for any $x_1, \ldots, x_n \in A$,

$$\mathcal{A} \models \exists y \varphi(y, x_1, \ldots, x_n) \Rightarrow \mathcal{A} \models \varphi(f(x_1, \ldots, x_n), x_1, \ldots, x_n),$$

i.e., f gives witnesses for the existential formulas. If for every φ of $\mathcal{L}$ there is a Skolem function f_φ then the closure of the set $\{f_\varphi \,;\, \varphi \text{ is a formula of } \mathcal{L}\}$ under function composition is a complete set of Skolem functions for $\mathcal{A}$. If $B \subseteq A$ then the closure of B under a complete set of Skolem functions for $\mathcal{A}$ is called the $\mathcal{A}$-Skolem hull of B, which we denote by

$$\mathsf{Hull}_{\mathcal{A}}(B).$$

By the Tarski-Vaught criterion, $\mathsf{Hull}_{\mathcal{A}}(B) \prec \mathcal{A}$.

LEMMA 0.23. (ZF) *Let $\mathcal{A}$ be a wellorderable structure with a wellorderable language L, and Y a set of elements of $\mathcal{A}$. Then* $|\mathsf{Hull}_{\mathcal{A}}(Y)| \leq |Y| + |L|$.

This is [**Hod97**, 1.2.3].

We won't expand on the basics any further here, but we will define what we need on top as we go along.

1

Symmetric forcing

In this chapter we mainly present symmetric forcing and give some examples of the resulting symmetric models. In the first section we will give a short forcing reminder, based on Kunen's *Set Theory* [**Kun80**]. We do that in order to fix our forcing notation which may vary in different sources.

In Section 2 we will describe first the technique of symmetric forcing with partial orders. This technique is described also in [**Jec03**, pages 249–261 of Chapter 15 and pages 221–223 of Chapter 14], in the first edition of this book [**Jec78**] in more detail, and fully in [**Jec73**]. In these sources this method is discussed for forcing with Boolean valued models. We won't look at the details of the Boolean algebra based version of this construction because it will be used only in Section 5 of Chapter 2. There, we will be following strictly the notation and theorems of Jech's presentation of Boolean valued models and symmetric forcing with Boolean algebras (see [**Jec03**, Chapters 14 and 15]). We will expand the background theory of symmetric forcing with partial orders by discussing partial orders that are $\mathcal{G}, I$-homogeneous. This is a property of the basic ingredients of symmetric forcing (the partial order, an automorphism group $\mathcal{G}$ of it, and a symmetry generator I) and it guarantees a resulting symmetric model whose sets of ordinals are in some inner generic extension that satisfies ZFC. Since we will be analysing properties of cardinals and of sets of ordinals in models of $\mathsf{ZF}+\neg\mathsf{AC}$, this is extremely useful in our study. Throughout this section we will be accompanied by a running example, the construction for the Feferman-Lévy model, to help the reader understand the definitions better. Details on this model can be found in [**Dim06**, §3.3].

In Section 3 we will go back to symmetric forcing to look at a generalisation of Jech's construction in ω_1 *can be measurable* [**Jec68**], giving the general construction and some specific examples, which involve some large cardinals becoming successor cardinals while retaining their large cardinal property.

Lastly, in Section 4 we will see how to modify the construction of Section 3 to get a model with an arbitrarily long sequence of cardinals which are alternating measurable and non-measurable cardinals. There the "measurable" can be replaced with other large cardinal properties that are preserved under mild symmetric forcing.

Before we start with each section, a few words on the history of symmetric models. The method of taking symmetric submodels of a generic extension is used to create models where the axiom of choice (AC) does not hold everywhere. This method is inspired by the older method of permutation models, by Fraenkel, Lindenbaum, and Mostowski, over 75 years ago. In that method, the ground model is a model of ZFCA, i.e., a modification of ZFC with the axiom for the existence of atoms added. Atoms have the same defining property of the empty set but are not equal to it. Therefore atoms are excluded from the axiom of extensionality. The axiom of the existence of atoms does not specify how many atoms are available, so these atoms can be as many as we want (by adding the axiom "the set of atoms is κ big") and they are indistinguishable from each other.

By permuting these atoms we can construct models of $\mathsf{ZFA} + \neg\mathsf{AC}$. The similarity between atoms in ZFCA and parts of the generic object in a forcing construction was noticed by several set theorists after the introduction of forcing. So the technique of permutation models was adapted to fit the forcing constructions and became the symmetric model technique. As we mentioned, the use of symmetry arguments goes back to Fraenkel [**Fra22a**]. A little later, Mostowski [**Mos39**], Lindenbaum [**LM38**], and Specker [**Spe57**] developed a general theory of this technique. Cohen used these arguments with his forcing technique [**Coh63**], [**Coh64**], and Jech [**Jec71**] and Scott (unpublished) formulated the technique in terms of Boolean valued models.

Roughly, to construct a symmetric model we use an automorphism group of the partial order and a normal filter over that group. We extend the automorphisms on the conditions to automorphisms on the names and in the new model we allow only these names that (hereditarily) remain intact under the permutations from a set in the filter. These names are called hereditarily symmetric and the class of their interpretations a symmetric model.

Even though the models of $\mathsf{ZF} + \neg\mathsf{AC}$ in this thesis are symmetric models, this technique is not the only way to obtain results for $\mathsf{ZF} + \neg\mathsf{AC}$. One can also use the method of taking $L(x)$ for certain x (see [**Jec03**, (13.24) in page 193]). This method takes all constructible sets definable from elements of x, without including a wellordering of x. Another similar method that is used is taking $\mathsf{HOD}(x)$ for some x (see Definition 3.43, and [**Jec03**, pages 195–196]).

1. Short forcing reminder and notation

Before we go into the details of symmetric models, we give a short reminder of the forcing technique and some words on the notation used here. A detailed introduction to forcing can be found in Kunen's *Set theory*, [**Kun80**, Chapter VII], and in Jech's *Set theory* [**Jec03**, Chapter 14]. Metamathematical concerns about forcing will not be addressed here, see [**Kun80**, Chapter VII].

Our background theory here is ZFC. Fix a model V of ZFC which we will call the ground model. A set $\mathbb{P}$ with a relation $\leq_{\mathbb{P}}$ is a partial order iff $\leq$ is reflexive and transitive. We will always assume that a partial order has a maximal element $1_{\mathbb{P}}$, and when we say $\mathbb{P}$ is partial order and mean $\langle \mathbb{P}, \leq_{\mathbb{P}}, 1_{\mathbb{P}} \rangle$ is a partial order. We will denote $\leq_{\mathbb{P}}$ and $1_{\mathbb{P}}$ by $\leq$ and 1 when it is clear from the context what we mean. So let $\mathbb{P} \in V$ be a partial order with maximal element 1. For two elements p, q of the partial order we say that p is stronger than q when $p \leq q$. As usual in forcing we call the elements of $\mathbb{P}$ conditions. We say that two conditions $p, q \in \mathbb{P}$ are compatible and we write $p \parallel q$ iff $\exists r \in \mathbb{P}(r \leq p \wedge r \leq q)$. If they are not compatible then we call them incompatible and we write $p \perp q$. A set $D \subseteq \mathbb{P}$ is dense in $\mathbb{P}$ if for every $p \in \mathbb{P}$ there is a $q \in D$ such that $q \leq p$.

To do a forcing construction we need to have a $\mathbb{P}$-generic filter over V, i.e., a filter G on the partial order that intersects all the dense subsets of $\mathbb{P}$ in V. We usually drop the "over V" when V is the ground model. In usual forcing constructions G is not in the ground model V (see [**Kun80**, Chapter VII, Lemma 2.4]). Intuitively to get a forcing extension we close $V \cup \{G\}$ under all set theoretic operations. Formally we take all objects that are definable in a particular way from G and finitely many elements of V. We do that by using names for the new objects, names that are elements of V.

Names are defined by recursion on the rank. A set of pairs τ is a $\mathbb{P}$-name iff τ is a relation and for every $(\sigma, p) \in \tau$, σ is a $\mathbb{P}$-name and $p \in \mathbb{P}$. We use both standard notations for names; $\dot{x}$ for a name of an already given x and σ, τ, etc. for arbitrary names. We write $V^{\mathbb{P}}$ for the class of $\mathbb{P}$-names in a ground model V.

We recursively define canonical names for objects in the ground model. For $x \in V$ the canonical name of x is

$$\check{x} \stackrel{\text{def}}{=} \{(\check{y}, \mathbb{1}) \,;\, y \in x\}.$$

Canonical names for sets of elements of V are names that consist of pairs $(\check{x}, p)$ for $p \in \mathbb{P}$. When G is a $\mathbb{P}$-generic filter over V and τ is a name, we write τ^G for the interpretation (valuation) of τ according to G, which is defined recursively as

$$\tau^G \stackrel{\text{def}}{=} \{\sigma^G \,;\, \exists p \in G((\sigma, p) \in \tau)\}.$$

The generic, or forcing extension is defined as

$$V[G] \stackrel{\text{def}}{=} \{\tau^G \,;\, \tau \in V^{\mathbb{P}}\},$$

it is a transitive model of ZFC, and it is the smallest ZFC model that contains both V and G (see [**Kun80**, Lemma 2.9].

DEFINITION 1.1. The forcing language $\mathcal{L}_F$ is a first order language that contains the $\in$-relation and all names as constants. For a formula $\varphi \in \mathcal{L}_F$ one could write φ^G for the φ with all its constants (names) evaluated in $V[G]$ and its unbounded quantifiers ranging over $V[G]$.

DEFINITION 1.2. Define the forcing relation $\Vdash$ between the conditions in $\mathbb{P}$ and the sentences of $\mathcal{L}_F$ as

$$p \Vdash \varphi(\vec{\tau}) \Longleftrightarrow \text{For every } V\text{-generic filter } G \text{ on } \mathbb{P} \text{ such that } p \in G,$$
$$V[G] \models \varphi^G(\vec{\tau}^G).$$

Note that this is a definition in our outer model, not in V. As one will read in [**Kun80**, Chapter VII, Definition 3.3 and Theorem 3.6], this $\Vdash$ is definable in V as well for a fixed formula φ of set theory, and the two definitions are equivalent. The following is referred to as *the forcing theorem.*

THEOREM 1.3. *Let $\langle \mathbb{P}, \leq \rangle$ be a partial order in the ground model V. If φ is a formula of $\mathcal{L}_F$ with n free variables then for every G that is a $\mathbb{P}$-generic filter and all $\mathbb{P}$-names $\tau_1, \ldots, \tau_n$*
,

$$V[G] \models \varphi^G(\tau_1^G, \ldots, \tau_n^G) \Longleftrightarrow \exists p \in G(p \Vdash \varphi(\tau_1, \ldots, \tau_n)).$$

This is Theorem 3.6 in [**Kun80**, Chapter VII]. Next we state the following useful properties of the forcing relation (See [**Jec78**, Exercises 16.7, 16.8, 16.9, and 16.10]).

PROPOSITION 1.4. *Let $\mathbb{P}$ be a partial order in V and φ, ψ arbitrary sentences in the forcing language for $\mathbb{P}$. Then the following hold.*

(a) *If $p \Vdash \varphi$ and $q \leq p$ then $q \Vdash \varphi$.*
(b) *There is no p such that $p \Vdash \varphi$ and $p \Vdash \neg\varphi$.*
(c) *For every p there is a $q \leq p$ such that $q \Vdash \varphi$ or $q \Vdash \neg\varphi$ (we say q decides φ).*
(d) *$p \Vdash \neg\varphi \Longleftrightarrow$ there is no $q \leq p$ such that $q \Vdash \varphi$.*
(e) *$p \Vdash \varphi \wedge \psi \Longleftrightarrow p \Vdash \varphi$ and $p \Vdash \psi$.*
$p \Vdash \forall x\varphi \Longleftrightarrow$ for every $\tau \in V^{\mathbb{P}}$, $p \Vdash \varphi(\tau)$.
(f) *$p \Vdash \varphi \vee \psi \Longleftrightarrow \forall q \leq p \exists r \leq q(r \Vdash \varphi$ or $r \Vdash \psi)$.*
$p \Vdash \exists x\varphi \Longleftrightarrow \forall q \leq p \exists r \leq q \exists \dot{a} \in V^{\mathbb{P}}(r \Vdash \varphi(\dot{a}))$.

We are not always going to make a distinction between φ and φ^G, since it will be clear from the context which one we mean.

We say that a partial order $\mathbb{P}$ preserves a cardinal κ of V iff for every $\mathbb{P}$-generic filter G, κ is a cardinal in $V[G]$. Similarly we say that a partial order preserves the cofinality of a limit ordinal γ of V iff for every $\mathbb{P}$-generic filter G, $(\mathsf{cf}(\gamma))^V = (\mathsf{cf}(\gamma))^{(G)}$. Two important facts about forcing are the following.

LEMMA 1.5. *Let ρ be a regular cardinal and $\mathbb{P}$ a partial order.*

If $\mathbb{P}$ has the ρ-chain condition (ρ-cc), i.e., of all its antichains have cardinality $< \rho$, then $\mathbb{P}$ preserves all cardinals and cofinalities $\geq \rho$ [**Kun80**, Chapter VII, Lemma 6.9].

If $\mathbb{P}$ is ρ-closed, i.e., if for every $\gamma < \rho$ and every γ-long decreasing sequence $\langle p_\alpha ; \alpha < \gamma\rangle$ of conditions in $\mathbb{P}$, there is a condition $p \in \mathbb{P}$ stronger than all the p_α's, then $\mathbb{P}$ preserves cardinals and cofinalities $\leq \rho$ [**Kun80**, Chapter VII, Lemma 6.15]

It's common to force with partial functions.

DEFINITION 1.6. Let X, Y be sets and λ a cardinal. Consider the partial order

$$\mathsf{Fn}(X, Y, \lambda) \stackrel{\text{def}}{=} \{p : X \rightharpoonup Y \ ; |p| < \lambda\},$$

partially ordered by reverse inclusion, i.e., $p \leq q :\Longleftrightarrow p \supseteq q$.
When $\lambda = \omega$ we write $\mathsf{Fn}(X, Y) \stackrel{\text{def}}{=} \mathsf{Fn}(X, Y, \omega)$.

This forcing, described in detail in [**Kun80**, Chapter VII, §6], adds a surjection from $|X|$ to $|Y|$ (see [**Kun80**, Chapter VII, Lemma 6.2]). By [**Kun80**, Chapter VII, Lemma 6.10] this forcing has the $(|Y|^{<\lambda})^+$-cc, and by [**Kun80**, Chapter VII, Lemma 6.13] if λ is regular then the forcing is λ-closed. This requirement for λ to be regular is the reason why we can't use this collapse when we want to collapse some ordinal to a singular cardinal. The topic of singular cardinals and collapsing onto them will be addressed in the next chapter.

Before we go on describing symmetric forcing we need some basic definitions and facts about complete and dense embeddings, product forcing, and iterated forcing.

DEFINITION 1.7. Let $\langle \mathbb{P}, \leq_{\mathbb{P}}, 1_{\mathbb{P}}\rangle$ and $\langle \mathbb{Q}, \leq_{\mathbb{Q}}, 1_{\mathbb{Q}}\rangle$ be partial orders, and let $i : \mathbb{P} \to \mathbb{Q}$.
We say that i is a complete embedding iff

(1) $\forall p, q \in \mathbb{P}(q \leq_{\mathbb{P}} p \Rightarrow i(q) \leq_{\mathbb{Q}} i(p))$,
(2) $\forall p, q \in \mathbb{P}(q \perp_{\mathbb{P}} p \Leftrightarrow i(q) \perp_{\mathbb{Q}} i(p))$, and
(3) $\forall q \in \mathbb{Q} \exists p \in \mathbb{P} \forall p' \in \mathbb{P}(p' \leq p \Rightarrow i(p') \parallel_{\mathbb{Q}} q)$.

We say that i is a dense embedding iff

(1) $\forall p, q \in \mathbb{P}(q \leq_{\mathbb{P}} p \Rightarrow i(q) \leq_{\mathbb{Q}} i(p))$,
(2) $\forall p, q \in \mathbb{P}(q \perp_{\mathbb{P}} p \Rightarrow i(q) \perp_{\mathbb{Q}} i(p))$, and
(3) the set $i``\mathbb{P}$ is dense in $\mathbb{Q}$.

In [**Kun80**, Chapter VII, Theorem 7.5] we see that if $\mathbb{P}$, $\mathbb{Q}$ are partial orders and $i : \mathbb{P} \to \mathbb{Q}$ is a complete embedding, then for every $\mathbb{Q}$-generic filter G, $i^{-1}``G$ is a $\mathbb{P}$-generic filter and $V[i^{-1}``G] \subseteq V[G]$. Clearly every dense embedding is a complete embedding. We quote [**Kun80**, Chapter VII, Theorem 7.11] below, which says that if there is a dense embedding between two partial orders then they produce the same generic extensions.

LEMMA 1.8. *Suppose i, $\mathbb{P}$, and $\mathbb{Q}$ are in V, and $i : \mathbb{P} \to \mathbb{Q}$ is a dense embedding. If $G \subseteq \mathbb{P}$ then let*

$$\tilde{\imath}(G) \stackrel{def}{=} \{q \in \mathbb{Q} \ ; \exists p \in G(i(p) \leq_{\mathbb{Q}} q)\}.$$

Then we have the following.

(a) *If $H \subseteq \mathbb{Q}$ is $\mathbb{Q}$-generic then $i^{-1}(H)$ is $\mathbb{P}$-generic and $H = \tilde{\imath}(i^{-1}(H))$.*
(b) *If $G \subseteq \mathbb{P}$ is $\mathbb{P}$-generic then $\tilde{\imath}(G)$ is $\mathbb{Q}$-generic and $G = i^{-1}(\tilde{\imath}(G))$.*

(c) *In (a) or (b), if* $G = i^{-1}(H)$ *(or equivalently* $H = \tilde{\imath}(G)$*), then* $V[G] = V[H]$.

Next we define product forcing.

DEFINITION 1.9. If $\langle \mathbb{P}, \leq_{\mathbb{P}}, 1_{\mathbb{P}} \rangle$ and $\langle \mathbb{Q}, \leq_{\mathbb{Q}}, 1_{\mathbb{Q}} \rangle$ are partial orders then the product partial order

$$\langle \mathbb{P}, \leq_{\mathbb{P}}, 1_{\mathbb{P}} \rangle \times \langle \mathbb{Q}, \leq_{\mathbb{Q}}, 1_{\mathbb{Q}} \rangle = \langle \mathbb{P} \times \mathbb{Q}, \leq_{\mathbb{P}\times\mathbb{Q}}, 1_{\mathbb{P}\times\mathbb{Q}} \rangle$$

is defined by

$$(p, q) \leq (p', q') \overset{\text{def}}{\Longleftrightarrow} p \leq_{\mathbb{P}} p' \text{ and } q \leq_{\mathbb{Q}} q',$$

and $1_{\mathbb{P}\times\mathbb{Q}} \overset{\text{def}}{=} (1_{\mathbb{P}}, 1_{\mathbb{Q}})$. Define $i_{\mathbb{P}} : \mathbb{P} \to \mathbb{P} \times \mathbb{Q}$ by $i_{\mathbb{P}}(p) = (p, 1_{\mathbb{Q}})$ and $i_{\mathbb{Q}} : \mathbb{Q} \to \mathbb{P} \times \mathbb{Q}$ by $i_{\mathbb{Q}}(q) = (1_{\mathbb{P}}, q)$.

As usual we drop the subscripts when it is clear from the context. The next lemma is [**Kun80**, Chapter VII, Lemma 1.3] and it says that we can easily split a generic filter on a product of partial orders to the parts of the product.

LEMMA 1.10. *Suppose* $\mathbb{P}, \mathbb{Q}$ *are partial orders in* V *and* G *is* $\mathbb{P} \times \mathbb{Q}$*-generic over* V*. Then* $i_{\mathbb{P}}^{-1}(G)$ *is* $\mathbb{P}$*-generic,* $i_{\mathbb{Q}}^{-1}(G)$ *is* $\mathbb{Q}$*-generic, and* $G = i_{\mathbb{P}}^{-1}(G) \times i_{\mathbb{Q}}^{-1}(G)$.

The next lemma is [**Kun80**, Chapter VII, Theorem 1.4] and it shows that we can force with a product or with either one of the parts first and then the other, to get the same result.

LEMMA 1.11. *Suppose* $\mathbb{P} \in V$, $\mathbb{Q} \in V$ *are partial orders,* $G \subseteq \mathbb{P}$ *and* $H \subseteq \mathbb{Q}$*. Then the following are equivalent:*

(1) $G \times H$ *is* $\mathbb{P} \times \mathbb{Q}$*-generic over* V.
(2) G *is* $\mathbb{P}$*-generic over* V *and* H *is* $\mathbb{Q}$*-generic over* $V[G]$.
(3) H *is* $\mathbb{Q}$*-generic over* V *and* G *is* $\mathbb{P}$*-generic over* $V[H]$.

Moreover, if (1)-(3) hold, then $V[G \times H] = V[G][H] = V[H][G]$.

We can also have longer products, but often we take only ones with finite support.

DEFINITION 1.12. If α is an ordinal and for every $i \in \alpha$, $\mathbb{P}_i$ is a partial order then

$$\prod_{i\in\alpha}^{\mathsf{fin}} \mathbb{P}_i \overset{\text{def}}{=} \{\vec{p} \in \prod_{i\in\alpha} ; \exists e \subset \alpha (e \text{ is finite and } \forall i \in \alpha \setminus e (\vec{p}(i) = 1_{\mathbb{P}_i}))\}.$$

We usually look at elements of such a finite support product as finite sequences with $\mathsf{dom}(\vec{p}) = e$. We may do that in a forcing context because there is always a complete embedding from $\prod_{i\in e} \mathbb{P}_i$ to $\prod_{i\in\alpha}^{\mathsf{fin}} \mathbb{P}_i$.

We end this section with the definition of two-stage iterated forcing.

DEFINITION 1.13. If $\mathbb{P}$ is a partial order, a $\mathbb{P}$-name for a partial order is a triple of $\mathbb{P}$-names, $\left\langle \dot{\mathbb{Q}}, \dot{\mathbb{Q}}', \dot{\mathbb{Q}}'' \right\rangle$ such that $\dot{\mathbb{Q}}'' \in \mathsf{dom}(\dot{\mathbb{Q}})$ and

$$1_{\mathbb{P}} \Vdash_{\mathbb{P}} (\dot{\mathbb{Q}}'' \in \dot{\mathbb{Q}} \text{ and } \dot{\mathbb{Q}}' \text{ is a partial order of } \dot{\mathbb{Q}} \text{ with largest element } \dot{\mathbb{Q}}'').$$

We write $\leq_{\dot{\mathbb{Q}}}$ instead of $\dot{\mathbb{Q}}'$ and $1_{\dot{\mathbb{Q}}}$ instead of $\dot{\mathbb{Q}}''$.

We then define $\mathbb{P} * \dot{\mathbb{Q}}$ to be the partial order whose base set is

$$\{\langle p, \tau\rangle \in \mathbb{P} \times \mathsf{dom}(\dot{\mathbb{Q}}) \; ; p \Vdash \tau \in \dot{\mathbb{Q}}\},$$

its partial ordering is defined by

$$\langle p, \tau\rangle \leq \langle q, \sigma\rangle \overset{\text{def}}{\iff} p \leq q \text{ and } p \Vdash \tau \leq_{\dot{\mathbb{Q}}} \sigma,$$

and its maximal element is $1_{\mathbb{P}*\dot{\mathbb{Q}}} \overset{\text{def}}{=} \langle 1_{\mathbb{P}}, 1_{\dot{\mathbb{Q}}}\rangle$.

2. The technique of symmetric forcing

We give a presentation of the technique of creating symmetric models in terms of forcing with partial orders. This is initially a translation to partial orders of the standard technique that Jech presents in his [**Jec03**] (for forcing with Boolean values). Moreover we introduce several notions that lead to the definition of a $\mathcal{G}, I$-homogeneity property for partial orders. Symmetric forcing with such a partial order ensures that in the resulting symmetric model the sets of ordinals are very well behaved.

As we define symmetric models we will have the following example in our minds. Let

$$\mathbb{F} \overset{\text{def}}{=} \{p : \omega \times \omega \rightharpoonup \aleph_\omega \; ; \; |p| < \omega \text{ and } \forall (n,i) \in \mathsf{dom}(p),\; p(n,i) < \omega_n\}$$

be ordered by reverse inclusion, i.e., $p \leq q$ iff $p \supseteq q$. This partial order is used to build the well known Feferman-Lévy model, first constructed in 1963 (for the abstract see [**FL63**] and for details see [**Dim06**, §3.3]). In this model, the reals are a countable union of countable sets and therefore both AC and AD fail.

It's easy to see that for every $n \in \omega$, the partial order $\mathbb{F}$ adds a countable set whose elements are surjective functions from ω to ω_n, i.e., a set of collapsing functions for ω_n.

In any generic extension the ordinal $\kappa = (\aleph_\omega)^V$ has become a countable union of countable sets and therefore is countable.

If $\langle \mathbb{P}, \leq, 1\rangle$ is a partial order, an automorphism a of $\mathbb{P}$ is a bijection of $\mathbb{P}$ to itself which preserves $\leq$ and 1 both ways. If a is an automorphism of $\mathbb{P}$, then define by recursion on $V^{\mathbb{P}}$,

$$a_*(\tau) \overset{\text{def}}{=} \{(a_*(\sigma), a(p)) \; ; (\sigma, p) \in \tau\}.$$

Given a, we will denote a_* also by a as it will be clear from the context what we mean. We will need to use an automorphism group $\mathcal{G}$ of our partial order.

For our running example consider $\mathcal{G}_{\mathbb{F}}$ to be the full permutation group of ω. Extend $\mathcal{G}_{\mathbb{F}}$ to an automorphism group of $\mathbb{F}$ by letting an $a \in \mathcal{G}_{\mathbb{F}}$ act on a $p \in \mathbb{F}$ by

$$a^*(p) \overset{\text{def}}{=} \{(n, a(i), \beta) \; ; (n, i, \beta) \in p\}.$$

We will identify a^* with $a \in \mathcal{G}_{\mathbb{F}}$. It's easy to check that this is indeed an automorphism group of $\mathbb{F}$.

The following is called *the symmetry lemma.*

LEMMA 1.14. *Let $\mathbb{P}$ be a partial order and $\mathcal{G}$ an automorphism group of $\mathbb{P}$. Let φ be a formula of set theory with n free variables and let $\tau_1, \ldots, \tau_n \in V^{\mathbb{P}}$ be names. If $a \in \mathcal{G}$ then*

$$p \Vdash \varphi(\tau_1, \ldots, \tau_n) \iff a(p) \Vdash \varphi(a(\tau_1), \ldots, a(\tau_n)).$$

This lemma is easy to prove by induction on the formula φ, using the properties of the forcing relation (Proposition 1.4).

DEFINITION 1.15. Let $\mathcal{G}$ be a group and $\mathcal{F} \subseteq \mathcal{P}(\mathcal{G})$ a filter over $\mathcal{G}$. We say that $\mathcal{F}$ is a normal filter if for every $K \in \mathcal{F}$ and every $a \in \mathcal{G}$, the conjugate aKa^{-1} is in $\mathcal{F}$.

In some sources, a filter over a group should always be a normal filter (e.g., in [**Jec03**, page 251, (15.34)]). We will always specify when a filter is normal.

In our running example, for every $n \in \omega$ define the following sets.

$$E_n \overset{\text{def}}{=} \{p \cap (n \times \omega \times \omega_n) \,;\, p \in \mathbb{F}\}$$
$$\mathsf{fix}E_n \overset{\text{def}}{=} \{a \in \mathcal{G}_{\mathbb{F}} \,;\, \forall p \in E_n(a(p) = p)\}, \text{ and}$$
$$\mathcal{F}_{\mathbb{F}} \overset{\text{def}}{=} \{X \subseteq \mathcal{G}_{\mathbb{F}} \,;\, \exists n \in \omega,\ \mathsf{fix}E_n \subseteq X\}.$$

The set $\mathcal{F}_{\mathbb{F}}$ is easily shown to be a normal filter over $\mathcal{G}_{\mathbb{F}}$.

For the rest of this section fix a partial order $\mathbb{P}$, an automorphism group $\mathcal{G}$ of $\mathbb{P}$, and a normal filter $\mathcal{F}$ over $\mathcal{G}$.

DEFINITION 1.16. For each $\tau \in V^{\mathbb{P}}$, we define its symmetry group with respect to $\mathcal{G}$ as

$$\mathsf{sym}^{\mathcal{G}}(\tau) \overset{\text{def}}{=} \{a \in \mathcal{G} \,;\, a(\tau) = \tau\}.$$

If we see $\mathcal{G}$ as an automorphism group of $V^{\mathbb{P}}$ then for a name τ, $\mathsf{sym}^{\mathcal{G}}(\tau)$ is, in algebraic terminology, the stabilizer group of τ. We say that τ is a symmetric name if $\mathsf{sym}^{\mathcal{G}}(\tau) \in \mathcal{F}$. We denote by $\mathsf{HS}^{\mathcal{F}}$ the class of all hereditarily symmetric names, i.e.,

$$\mathsf{HS}^{\mathcal{F}} \overset{\text{def}}{=} \{\tau \in V^{G} \,;\, \forall \sigma \in \mathsf{tc}_{\mathsf{dom}}(\tau)(\mathsf{sym}^{\mathcal{G}}(\sigma) \in \mathcal{F})\},$$

where $\mathsf{tc}_{\mathsf{dom}}(\tau)$ is defined as the union of all x_n, which are defined recursively by $x_0 \overset{\text{def}}{=} \{\tau\}$ and $x_{n+1} \overset{\text{def}}{=} \bigcup\{\mathsf{dom}(\sigma) \,;\, \sigma \in x_n\}$.

When it is clear from the context we will denote these notions by sym, HS. By induction we can prove the next lemma which says that all canonical names are hereditarily symmetric.

LEMMA 1.17. *If a is an automorphism of $\mathbb{P}$ then for every canonical name $\check{x} \in V^{\mathbb{P}}$, $a(\check{x}) = \check{x}$.*

DEFINITION 1.18. We define the symmetric model with respect to $\mathcal{F}$,G by

$$V(G)^{\mathcal{F}} \stackrel{\text{def}}{=} \{\tau^G \; ; \tau \in \mathsf{HS}^{\mathcal{F}}\}.$$

We will often denote a symmetric model by simply $V(G)$. To talk about the truth of formulas in a symmetric model we use the symmetric forcing relation.

DEFINITION 1.19. For a formula φ and names $\vec{\tau}$ in HS, we informally define the relation $\Vdash_{\mathsf{HS}}$ by

$$p \Vdash_{\mathsf{HS}} \varphi(\vec{\tau}) \stackrel{\text{def}}{\Longleftrightarrow} \text{ for any } \mathbb{P}\text{-generic filter } G, \text{ and any } p \in G,\ V(G) \models \varphi$$

This symmetric forcing relation can be formally defined in the ground model similarly to the usual forcing relation $\Vdash$, and it has the some of the same properties as $\Vdash$ but with the quantifiers ranging over symmetric names. In particular we have the following proposition.

PROPOSITION 1.20. *If φ and ψ are arbitrary sentences in the forcing language for $\mathbb{P}$ then the following hold.*

(a) *If $p \Vdash_{\mathsf{HS}} \varphi$ and $p \leq q$ then $q \Vdash_{\mathsf{HS}} \varphi$.*
(b) *There is no p such that $p \Vdash_{\mathsf{HS}} \varphi$ and $p \Vdash_{\mathsf{HS}} \neg\varphi$.*
(c) *For every p there is a $q \leq p$ such that $q \Vdash_{\mathsf{HS}} \varphi$ or $q \Vdash_{\mathsf{HS}} \neg\varphi$.*
(d) *$p \Vdash_{\mathsf{HS}} \neg\varphi \iff$ there is no $q \leq p$ such that $q \Vdash_{\mathsf{HS}} \varphi$.*
(e) *$p \Vdash_{\mathsf{HS}} \varphi \wedge \psi \iff p \Vdash_{\mathsf{HS}} \varphi$ and $p \Vdash_{\mathsf{HS}} \psi$.*
$p \Vdash_{\mathsf{HS}} \forall x\varphi \iff$ for every $\tau \in \mathsf{HS}$, $p \Vdash_{\mathsf{HS}} \varphi(\tau)$.
(f) *$p \Vdash_{\mathsf{HS}} \varphi \vee \psi \iff \forall q \leq p \exists r \leq q (r \Vdash_{\mathsf{HS}} \varphi$ or $r \Vdash_{\mathsf{HS}} \psi)$.*
$p \Vdash_{\mathsf{HS}} \exists x\varphi \iff \forall q \leq p \exists r \leq q \exists \dot{a} \in \mathsf{HS}(r \Vdash_{\mathsf{HS}} \varphi(\dot{a}))$.

THEOREM 1.21. *A symmetric model $V(G)^{\mathcal{F}}$ is a transitive model of* ZF *and $V \subseteq V(G)^{\mathcal{F}} \subseteq V[G]$.*

PROOF. That $V \subseteq V(G)^{\mathcal{F}} \subseteq V[G]$ is obvious and by the heredity of the elements of HS, we get that $V(G)^{\mathcal{F}}$ is transitive. Extensionality, foundation, empty set, and infinity hold because $V \subseteq V(G)^{\mathcal{F}}$ and $V(G)^{\mathcal{F}}$ is transitive. For the separation schema let φ be a formula and let $y = \{x \in z \; ; V(G)^{\mathcal{F}} \models \varphi(x, w)\}$ where $z, w \in V(G)^{\mathcal{F}}$ with names $\dot{z}, \dot{w} \in \mathsf{HS}$ respectively. Define a name for y by

$$\dot{y} \stackrel{\text{def}}{=} \{(\sigma, p) \; ; \sigma \in \mathsf{dom}(\dot{z}) \text{ and } p \Vdash_{\mathsf{HS}} \varphi(\sigma, \dot{w})\},$$

This is a HS name for y because for every $a \in \mathsf{sym}(\dot{z}) \cap \mathsf{sym}(\dot{w})$, we have that

$$\begin{aligned} a(\dot{y}) &= \{(a(\sigma), a(p)) \; ; a(\sigma) \in \mathsf{dom}(\dot{z}) \text{ and } a(p) \Vdash_{\mathsf{HS}} \varphi(a(\sigma), \dot{w})\} \\ &= \{(\tau, q) \; ; a^{-1}(\tau) \in \mathsf{dom}(\dot{z}) \text{ and } a^{-1}(q) \Vdash_{\mathsf{HS}} \varphi(a^{-1}(\sigma), \dot{w})\} \\ &= \{(\tau, q) \; ; \tau \in \mathsf{dom}(\dot{z}) \text{ and } q \Vdash_{\mathsf{HS}} \varphi(\sigma, \dot{w})\} = \dot{y} \end{aligned}$$

Thus $y \in V(G)^{\mathcal{F}}$ and separation holds.

Now let $x \in V(G)^{\mathcal{F}}$ and let $\dot{x} \in \mathsf{HS}$ be a name for x. For the union axiom take $\tau \overset{\text{def}}{=} \{(\sigma, 1) ; \exists\pi \in \mathsf{dom}(\dot{x})(\sigma \in \mathsf{dom}(\pi))\}$ and remember that if $a \in \mathsf{sym}(\dot{x})$ then $a(\dot{x}) = \dot{x}$. This means that the names in $\mathsf{dom}(\dot{x})$ may be permuted with each other but overall $\dot{x}$ stays the same; thus $a(\tau) = \tau$ as well. Clearly $\tau^G \supseteq \bigcup x$ holds and so because of separation we have that union also holds.

For pairing of $x, y \in V(G)^{\mathcal{F}}$ with names $\dot{x}, \dot{y} \in \mathsf{HS}$ respectively, we take $\tau \overset{\text{def}}{=} \{(\dot{x}, 1), (\dot{y}, 1)\}$. For the powerset of x we take the name $\sigma \overset{\text{def}}{=} \{(\pi, 1) ; \pi \in \mathsf{HS} \text{ and } \mathsf{dom}(\pi) \subset \mathsf{dom}(\dot{x})\}$. This σ is in HS and gives a superset of the powerset of x. So using separation we get that $V(G)^{\mathcal{F}}$ satisfies the powerset axiom.

Finally we do replacement. We want to show that if $x \in V(G)^{\mathcal{F}}$ with $\dot{x} \in \mathsf{HS}$ a name for it and φ is a function-like formula, then there is a $y \in V(G)^{\mathcal{F}}$ such that

$$V(G)^{\mathcal{F}} \models (\forall z \in x \exists w(\varphi(z, w)) \Rightarrow \forall z \in x \exists w \in y(\varphi(z, w))).$$

For every $\dot{z} \in \mathsf{dom}(\dot{x})$ define $S_{\dot{z}} \overset{\text{def}}{=} \{\dot{w} \in \mathsf{HS} ; \exists p \in \mathbb{P}(\ p \Vdash_{\mathsf{HS}} \varphi(\dot{z}, \dot{w}))\}$ and

$$\tau \overset{\text{def}}{=} \{(\sigma, 1) ; \exists \dot{z} \in \mathsf{dom}(\dot{x})(\sigma \in T_{\dot{z}})\}.$$

Now use separation in $V(G)$ to get the $y \subseteq \tau^G$ we were looking for. qed

In the older method of permutation models it is common to build the normal filter via a normal ideal. The following notion of a symmetry generator corresponds to the notion of a normal ideal.

DEFINITION 1.22. For an $E \subseteq \mathcal{P}$ we take its pointwise stabilizer group, i.e.,

$$\mathsf{fix}_{\mathcal{G}} E \overset{\text{def}}{=} \{a \in \mathcal{G} ; \forall p \in E(a(p) = p)\}$$

which is the set of automorphisms that do not move the elements of E (they fix E). Usually we will just write $\mathsf{fix}E$.

Call $I \subseteq \mathcal{P}(\mathbb{P})$ a $\mathcal{G}$-symmetry generator if it is closed under taking unions and if for all $a \in \mathcal{G}$ and $E \in I$, there is an $E' \in I$ such that $a(\mathsf{fix}E)a^{-1} \supseteq \mathsf{fix}E'$.

PROPOSITION 1.23. *If I is a $\mathcal{G}$-symmetry generator then the set $\{\mathsf{fix}E ; E \in I\}$ generates a normal filter $\mathcal{F}_I$ over $\mathcal{G}$.*

PROOF. Let $K_1, K_2 \in \mathcal{F}_I$. There are $E_1, E_2 \in I$ such that $\mathsf{fix}E_1 \subseteq K_1$ and $\mathsf{fix}E_2 \subseteq K_2$. Then $\mathsf{fix}(E_1 \cup E_2) = \mathsf{fix}E_1 \cap \mathsf{fix}E_2 \subseteq K_1 \cap K_2$ and $E_1 \cup E_2 \in I$. So $K_1 \cap K_2 \in \mathcal{F}_I$.

Because $\mathcal{F}_I$ is being generated we immediately get that the rest of the filter axioms hold as well. For normality let $a \in \mathcal{G}$ and $K \in \mathcal{F}_I$. Then there is $E \in I$ such that $\mathsf{fix}E \subseteq K$. Then $a(\mathsf{fix}E)a^{-1} \subseteq aKa^{-1}$ and there is $E' \in I$ such that $\mathsf{fix}E' \subseteq a\mathsf{fix}Ea^{-1}$. So $aKa^{-1} \in \mathcal{F}_I$. qed

DEFINITION 1.24. We say that a set $E \in I$ supports a name $\sigma \in \mathsf{HS}$ if $\mathsf{sym}(\sigma) \supseteq \mathsf{fix}E$.

Constructing symmetric models with symmetry generators also helps describe a very nice property of sets of ordinals.

2.1. The approximation lemma. We can have a lot of grip on what a symmetric model construction does to the sets of ordinals if we could describe these (wellorderable) sets in some inner ZFC model of $V(G)$. In particular to know them by knowing only an initial part of the forcing construction. This is exactly what the approximation property guarantees. Before we go on to say what this property is let us take a look at the symmetry generators we must use to describe it.

DEFINITION 1.25. Let $\mathbb{P}$ be a partial order and $\mathcal{G}$ an automorphism group of $\mathbb{P}$. A symmetry generator I is called projectable for $\mathbb{P}, \mathcal{G}$ if for every $p \in \mathbb{P}$ and every $E \in I$, there is a $p^* \in E$ that is minimal (with respect to the partial order) and unique such that $p^* \geq p$. Call this $p^* = p \restriction^* E$ the projection of p to E.

We are going to use only projectable I's. They will be comprised by either initial segments of the partial order (like with the Feferman-Lévy model) or by chunks of the partial order (like in the case of the products in Chapter 2).

In our running example we take the symmetry generator $L \stackrel{\text{def}}{=} \{E_n \,;\, n \in \omega\}$.
It's easy to see that L is a projectable symmetry generator with projections
$p{\restriction^*}E_n = p \cap (n \times \omega \times \omega_n)$.

Now we can describe the property of $\mathbb{P}$ being $\mathcal{G}, I$-homogeneous.

DEFINITION 1.26. Let $\mathbb{P}$ be a partial order, $\mathcal{G}$ an automorphism group of $\mathbb{P}$, and I be a projectable symmetry generator for $\mathbb{P}$,$\mathcal{G}$. We say that $\mathbb{P}$ is $\mathcal{G}, I$-homogeneous if for every $E \in I$, every $p \in \mathbb{P}$, and every $q \in \mathbb{P}$ such that $q \leq p \restriction^* E$, there is an automorphism $a \in \mathsf{fix}E$ such that $ap \parallel q$,

Next we see a useful consequence of $\mathcal{G}, I$-homogeneneity.

LEMMA 1.27. *Let $\mathbb{P}$ be a partial order, $\mathcal{G}$ an automorphism group of $\mathbb{P}$, and I be a projectable symmetry generator for $\mathbb{P}, \mathcal{G}$. If $\mathbb{P}$ is $\mathcal{G}, I$-homogeneous, then for any formula φ with n-many free variables, any names $\sigma_1, \ldots, \sigma_n \in \mathsf{HS}$ all with support $E \in I$, and any $p \in \mathbb{P}$,*

$$p \Vdash \varphi(\sigma_1, \ldots, \sigma_n) \textit{ implies that } p \restriction^* E \Vdash \varphi(\sigma_1, \ldots, \sigma_n).$$

PROOF. Assume the contrary, i.e., that $p{\restriction^*}E \nVdash \varphi(\sigma_1, \ldots, \sigma_n)$. Then there is a $q \leq p{\restriction^*}E$ such that $q \Vdash \neg\varphi(\sigma_1, \ldots, \sigma_n)$. Take $a \in \mathsf{fix}E$ such that $a(p) \parallel q$. By the symmetry lemma, $a(p) \Vdash \varphi(a(\sigma_1), \ldots, a(\sigma_n))$. Because $a \in \mathsf{fix}E$ and E supports $\sigma_1, \ldots, \sigma_n$ we get $a(p) \Vdash \varphi(\sigma_1, \ldots, \sigma_n)$ which is a contradiction. qed

LEMMA 1.28. *In our running example $\mathbb{F}$ is $\mathcal{G}_{\mathbb{F}}, L$-homogeneous.*

PROOF. Assume $E_n \in L$ and $q \in \mathbb{F}$ is such that $q \leq p{\restriction^*}E_n$.
Let $(\ )_1$ denote projection to the first coordinate. Since p is finite there is a set $A \subseteq \omega \setminus ((p)_1 \cup (q)_1 \cup n)$ that is equinumerous to $(p)_1 \setminus n$. Let

$f : (p)_1 \setminus n \to A$ be an injection and define a permutation a of ω by

$$a(m) \stackrel{\text{def}}{=} \begin{cases} m & \text{if } m \notin ((p)_1 \setminus n) \cup A \\ f(m) & \text{if } m \in (p)_1 \setminus n \\ f^{-1}(m) & \text{if } m \in A \end{cases}$$

as depicted below.

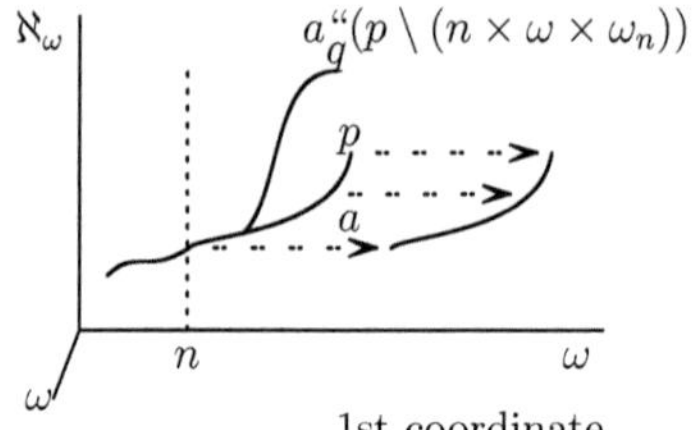

This defines also an automorphism of $\mathbb{F}$ which we denote also by a. Clearly $a \in \mathsf{fix}E_n$ and $a(p) \parallel q$, so $\mathbb{F}$ is $\mathcal{G}_{\mathbb{F}}, L$-homogeneous. qed

The following is called the approximation lemma. It's a very useful way of describing sets of ordinals in symmetric models that have been constructed with partial orders that are $\mathcal{G}, I$-homogeneous.

LEMMA 1.29. *If $\mathbb{P}$ is a partial order, $\mathcal{G}$ is an automorphism group, and I a projectable symmetry generator such that $\mathbb{P}$ is $\mathcal{G}, I$-homogeneous, then for any set of ordinals $X \in V(G)^{\mathcal{F}_I}$ there is an $E \in I$ and an E-name for X, therefore*

$$X \in V[G \cap E].$$

PROOF. Let $\dot{X} \in \mathsf{HS}$ be a name for X with support $E \in I$. Define a name

$$\ddot{X} \stackrel{\text{def}}{=} \{(\check{\alpha}, p\restriction^* E) \,;\, p \Vdash \check{\alpha} \in \dot{X}\}.$$

Because $\mathbb{P}$ is $\mathcal{G}, I$-homogeneous, this is a name for X. qed

For more details on the Feferman-Lévy model, see [**Dim06**, §3.3], where $V(G)^{\mathcal{F}_L}$ is denoted by $\mathcal{M}9$. There it is shown that the ordinal $(\omega_\omega)^V$ is a cardinal in $V(G)^{\mathcal{F}_L}$, the ordinal $(\omega_1)^{V(G)^{\mathcal{F}_L}}$ is singular in $V(G)^{\mathcal{F}_L}$, for every $n \in \omega$, $(\omega_{n+2})^{V(G)^{\mathcal{F}_L}} = (\omega_{n+1})^{V[G]}$, and that the set of all reals in $V(G)^{\mathcal{F}_L}$ is a countable union of countable sets.

3. Large cardinals that are successor cardinals

We now turn back to symmetric forcing. In 1968 Thomas Jech published a paper in which ω_1 was made measurable [**Jec68**], thus showing that the axiom of choice is necessary in proving that measurable cardinals are limit cardinals (under AC they are in fact inaccessible). In this section we will give a generalised version of this by constructing a model where any desired successor may become measurable, and extend this to other large cardinal properties.

Jech's classic symmetric model construction is done with injective partial functions. We found that this requirement of injectivity is not necessary and we present this construction with the standard forcing $\mathsf{Fn}(\eta, \kappa, \eta)$. First we make it into a passe-partout construction, using an inaccessible cardinal κ, and a regular cardinal $\eta < \kappa$. Then we discuss some large cardinal properties that imply inaccessibility (weak compactness, being Ramsey, etc.) and how they can be implemented in this construction. This way we will be able to easily use this construction for many results. The original construction from Jech can be found in [**Jec78**, page 476] and [**Jec68**].

3.1. The general construction. Assume that we are in a model of ZFC in which there is an inaccessible cardinal κ and a regular cardinal $\eta < \kappa$. Note that here we can have $\eta = \omega$, or $\eta = \aleph_{\omega+1}$, etc.. Let

$$\mathbb{P} \stackrel{\text{def}}{=} \mathsf{Fn}(\eta, \kappa, \eta) = \{p : \eta \rightharpoonup \kappa \,;\, |p| < \eta\}.$$

Let $\mathcal{G}$ be the full permutation group of κ and extend it to the partial order by permuting the range of the conditions, i.e., for $a \in \mathcal{G}$ and $p \in \mathbb{P}$,

$$a(p) \stackrel{\text{def}}{=} \{(\xi, a(\beta)) \,;\, (\xi, \beta) \in p\}.$$

Let I be the following symmetry generator.

$$I \stackrel{\text{def}}{=} \{E_\alpha \,;\, \eta < \alpha < \kappa\}, \text{ where}$$
$$E_\alpha \stackrel{\text{def}}{=} \{p \cap (\eta \times \alpha) \,;\, p \in \mathbb{P}\}.$$

It is clear that this is a projectable symmetry generator with projections $p{\restriction}^* E_\alpha = p \cap (\eta \times \alpha)$. Take the symmetric model $V(G) = V(G)^{\mathcal{F}_I}$. Similarly to Lemma 1.28 we can prove the following.

Lemma 1.30. *The partial order $\mathbb{P}$ is $\mathcal{G}, I$-homogeneous, i.e., for every $\alpha \in (\eta, \kappa)$, every $p \in \mathbb{P}$ and every $q \in \mathbb{P}$ such that $q \leq p{\restriction}^* E_\alpha$, there is an automorphism $a \in \mathsf{fix}(E_\alpha)$ such that $a(p) \parallel q$. Consequently, the approximation lemma holds for $V(G)$.*

In this model κ has become η^+.

Proposition 1.31. *In $V(G)$, κ is the successor of η.*

Proof. Let $\gamma < \kappa$ and define the following name.

$$\tau \stackrel{\text{def}}{=} \{(\check{p}, p) \,;\, p \in \mathbb{P}{\restriction}^* E_\gamma\}.$$

This is a hereditarily symmetric $\mathbb{P}$-name (supported by γ) and $\bigcup \tau_G$ is a surjection of η onto γ.

Assume towards a contradiction that there is some $\beta < \kappa$ and a bijection $f : \beta \to \kappa$ in $V(G)$. Let $\dot{f} \in \mathsf{HS}$ be a name for f with support E_δ. By the approximation lemma, $f \in V[G \cap E_\delta]$ which is impossible because κ is inaccessible and since E_δ has cardinality $< \kappa$, it has the κ-cc. qed

By [**Kun80**, Chapter VII, Lemma 6.10], $\mathbb{P}$ has the $(|\kappa|^{<\eta})^+$-cc, and since κ is inaccessible, $\mathbb{P}$ has the κ^+-cc, so all cardinals above κ are also preserved. Moreover, since η is regular, by [**Kun80**, Chapter VII, Lemma 6.13], $\mathbb{P}$ is η-closed, so all cardinals $\leq \eta$ are preserved as well.

This preservation of cardinals is the only place where the inaccessibility of κ is used. If κ weren't inaccessible we could require that κ is a limit cardinal and that GCH holds below κ, to get the same results.

LEMMA 1.32. *In $V(G)$, the powerset of η is a κ-sized union of sets with size $\leq \kappa$.*

PROOF. Let $\tau \in \mathsf{HS}$ be a name for $(\mathcal{P}(\eta))^{V(G)}$ whose domain contains only nice names for subsets of η. For every ordinal α such that $\eta < \alpha < \kappa$ define

$$C_\alpha \stackrel{\text{def}}{=} \{\dot{x}^G \subseteq \eta \,;\, E_\alpha \text{ supports } \dot{x} \text{ and } \dot{x} \in \mathsf{dom}(\tau)\}.$$

Clearly, $(\mathcal{P}(\eta) = \bigcup_{\eta<\alpha<\kappa} C_\alpha)^{V(G)}$.

For every $\eta < \alpha < \kappa$ and every $x \in C_\alpha$ define

$$\ddot{x} \stackrel{\text{def}}{=} \{(\check{\xi}, p\restriction^* E_\alpha) \,;\, p \Vdash \check{\xi} \in \dot{x}\}, \text{ and}$$
$$C'_\alpha \stackrel{\text{def}}{=} \{\ddot{x} \,;\, \dot{x} \in C_\alpha\}.$$

By the approximation lemma we have that $\dot{x}^G = \ddot{x}^G$, so it suffices to show that C'_α has cardinality less than κ in $V(G)$. In V we have that C'_α injects into $(\mathcal{P}^{(k)}(\alpha))^V$ for some finite k. But in V there is a bijection from $(\mathcal{P}^{(k)}(\alpha))^V$ to some cardinal $\xi_\alpha < \kappa$ because κ is inaccessible in V. So we can get an injection in $V(G)$ from C_α to $\xi_\alpha < \kappa$ for each $\eta < \alpha < \kappa$. qed

3.2. Measurability. Here we give Jech's result that the theory "ZF + ω_1 is measurable" is consistent relative to "ZFC + there exists a measurable". Our construction uses partial orders but one can see the Boolean valued model argumentation in [**Jec78**, page 476] and [**Jec68**].

Remember that by "κ is measurable" here we mean that there is a κ-complete non-trivial ultrafilter U over κ. To show measurability of ω_1 or any other successor cardinal in the symmetric model that we will construct, we will use the Lévy-Solovay theorem [**LS67**] which says that measurability is preserved under small forcing (forcing of cardinality less than the measurable). In particular it says that if U is a measure for the measurable cardinal then the set generated from U by taking supersets is still a measure after small forcing.

LEMMA 1.33. *If κ is a measurable cardinal and $\eta < \kappa$ is a regular cardinal then there is a symmetric model in which η^+ is measurable.*

Note that this lemma gives an infinity of consistency strength results, by replacing η with a description for a regular cardinal such as "η is ω_1", etc..

PROOF. Take $V(G)$ to be Jech's construction with the same notation as in the previous section, and let U be a κ-complete ultrafilter over κ in V. In $V(G)$ define the set

$$W \stackrel{\text{def}}{=} \{x \subseteq \kappa \,;\, \exists y \in U, y \subseteq x\}.$$

This is clearly a filter in $V(G)$ so it remains to show that it's also a κ-complete ultrafilter. For this we need to use the approximation lemma. To show that it is an ultrafilter, let $X \subseteq \kappa$, $X \in V(G)$, and let $\dot{X} \in \mathsf{HS}$ be a name for X, supported by $E_e \in I$. By the approximation lemma we have that $X \in V[G \cap E_e]$ so we can use the Lévy-Solovay theorem to show that either $X \in W$ or $\kappa \setminus X \in W$.

For the κ-completeness of W we can't use that W is generated by a κ-complete filter because $V(G)$ does not satisfy AC. So to show that it is κ-complete let $\gamma < \kappa$ and $\langle X_\delta \ ; \ \delta < \gamma \rangle$ be a sequence of sets in W. Let $\sigma \in \mathsf{HS}$ be a name for this sequence and let $E_f \in I$ be a support for this sequence. Since a sequence of sets of ordinals can be coded into a set of ordinals, we can use the approximation lemma to get that the sequence is in $V[G \cap E_f]$. Again by the Lévy-Solovay theorem we get that its intersection is in W. Therefore W is a measure for η in $V(G)$.

Therefore κ which now is the η^+ of $V(G)$, is measurable in $V(G)$. qed

By Corollary 0.13, in this and in the next model AC_κ fails.

As pointed out by Lorenz Halbeisen, in this construction κ has a normal measure as well: Assume that in the proof above W is a normal ultrafilter in the ground model. Then any regressive function can be coded as a set of ordinals, thus is in some $V[G \cap E_e]$ and by the Lévy-Solovay Theorem again, it is constant in some set in W.

That left the author wandering whether there exists a model of ZF in which there is a measurable cardinal without a normal measure. During the final stages of the thesis the author was thankful to receive from Moti Gitik a preprint of a paper with Eilon Bilinsky in which they indeed construct such a model.

3.3. Generalising to other large cardinal properties. The next lemma is mentioned as provable in [**Jec68**, pages 366-367].

LEMMA 1.34. *If κ is a weakly compact cardinal, i.e., a cardinal such that $\kappa \to (\kappa)^2_2$, and $\eta < \kappa$ is a regular cardinal then there is a symmetric model in which (η^+) is weakly compact, i.e., $\eta^+ \to (\eta^+)^2_2$.*

PROOF. Take $V(G)$ to be Jech's construction. We will show that $\kappa \to (\kappa)^2_2$ holds in $V(G)$. Let $f \colon [\kappa]^2 \to 2$ in $V(G)$, $\dot{f} \in \mathsf{HS}$ a name for f, and for some $\alpha < \kappa$ E_α is a support for $\dot{f}$. By the approximation lemma $f \in V[G \cap E_\alpha]$ and by [**Jec03**, Theorem 21.2], κ is still weakly compact in $V[G \cap E_\alpha]$. So in $V[G \cap E_\alpha] \subseteq V(G)$ there is a set $H \in [\kappa]^\kappa$ that is homogenous for f. qed

This proof is based on that weak compactness is preserved under small forcing, and on that we can pull our the function f premise in an inner model of ZFC. So we can replace "measurable" or "weakly compact" by some large cardinal property that is preserved under small forcing, and which is of the form

"for every set of ordinals X, there is a set Y such that $\varphi(X, Y)$ holds" for downwards absolute formulas φ with two free variables. This is because for such properties we can use the approximation lemma to capture the arbitrary set of ordinals in an intermediate ZFC model that is included in the symmetric model and use small forcing arguments to prove that such a large cardinal property is preserved. This allows us to construct models in which, e.g., a Ramsey cardinal is a successor cardinal, starting from a model of ZFC+"there exists a Ramsey cardinal".

When a large cardinal notion has this property above, we say that it is *preserved under small symmetric forcing.*

Note that the partition property $\kappa \to (\alpha)_2^{<\omega}$ is preserved under small symmetric forcing (with the same arguments as for the weakly compact) but Erdős cardinals are not necessarily preserved under small symmetric forcing. The requirement that they are minimal such that a partition property holds, may not be preserved under arbitrary small symmetric forcing.

4. Alternating measurable and non-measurable cardinals

In this section we will construct a sequence of cardinals, which are alternating measurable and non-measurable cardinals, when we exclude the singular limits. When we reach a singular limit the pattern will be "singular-regular-measurable", since collapsing a measurable to be the successor of a singular requires much more consistency strength than just ZFC. We will see such constructions in the next Chapter.

We start with a model of ZFC, an ordinal ρ, and a sequence $\langle \kappa_\xi\ ;\ \xi < \rho\rangle$ such that for every $0 < \xi < \rho$, κ_ξ is measurable, and κ_0 is any chosen regular cardinal. For each $\xi \in (0, \rho)$ define the following cardinals:

$$\kappa_1' \stackrel{\text{def}}{=} \kappa_0,$$

$$\kappa_\xi' \stackrel{\text{def}}{=} (\kappa_{\xi-1}^+)^V, \text{ if } \xi \text{ is a successor ordinal,}$$

$$\kappa_\xi' \stackrel{\text{def}}{=} ((\bigcup_{\zeta<\xi} \kappa_\zeta)^+)^V \text{ if } \xi \text{ is a limit ordinal and } \bigcup_{\zeta<\xi} \kappa_\zeta \text{ is singular,}$$

$$\kappa_\xi' \stackrel{\text{def}}{=} \bigcup_{\zeta<\xi} \kappa_\zeta \text{ if } \xi \text{ is a limit ordinal and } \bigcup_{\zeta<\xi} \kappa_\zeta < \kappa_\xi \text{ is regular, and}$$

$$\kappa_\xi' \stackrel{\text{def}}{=} (\bigcup_{\zeta<\xi} \kappa_\zeta)^{++} \text{ if } \xi \text{ is a limit ordinal and } \bigcup_{\zeta<\xi} \kappa_\zeta = \kappa_\xi \text{ is regular.}$$

For each $0 < \xi < \rho$ we will do the construction of the previous section between κ_ξ' and κ_ξ, as depicted in the picture below.

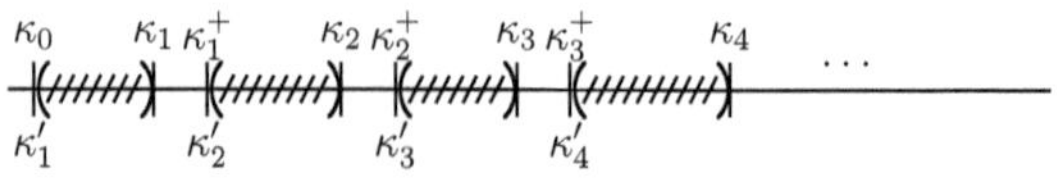

So for each $0 < \xi < \rho$ define

$$\mathbb{P}_\xi \stackrel{\text{def}}{=} \mathsf{Fn}(\kappa'_\xi, \kappa_\xi, \kappa'_\xi).$$

We will force with the full product of all these partial orders:

$$\mathbb{P} \stackrel{\text{def}}{=} \prod_{0<\xi<\rho} \mathbb{P}_\xi$$

For $\vec{p} \in \mathbb{P}$ and $\zeta \in (0, \rho)$, we sometimes write p_ζ instead of $\vec{p}(\zeta)$. For each $0 < \xi < \rho$ let $\mathcal{G}_\xi$ be the full permutation group of κ_ξ and extend it to $\mathbb{P}_\xi$ by permuting the range of its conditions, i.e., for $a \in \mathcal{G}_\xi$ and $p \in \mathbb{P}_\xi$ let

$$a(p) \stackrel{\text{def}}{=} \{(\zeta, a(\beta)) \,;\, (\zeta, \beta) \in p\}.$$

Take the product of these $\mathcal{G}_\xi$ to get an automorphism group $\mathcal{G}$ of $\mathbb{P}$:

$$\mathcal{G} \stackrel{\text{def}}{=} \prod_{\xi<\rho} \mathcal{G}_\xi.$$

For every finite sequence of ordinals $e = \langle \alpha_1, \ldots, \alpha_m \rangle$ such that for every $i = 1, \ldots, m$ there is a distinct $\zeta_i \in (0, \rho)$ such that $\alpha_i \in (\kappa'_{\zeta_i}, \kappa_{\zeta_i})$, define

$$E_e \stackrel{\text{def}}{=} \{\langle \ldots, p_{\zeta_1} \cap (\kappa'_{\zeta_1} \times \alpha_1), \varnothing, \ldots, p_{\zeta_i} \cap (\kappa'_{\zeta_i} \times \alpha_i), \varnothing, \ldots \rangle \,;\, \vec{p} \in \mathbb{P}\}, \text{ and}$$

$$I \stackrel{\text{def}}{=} \{E_e \,;\, e \in \prod_{\xi\in(0,\rho)}^{\mathsf{fin}} (\kappa'_\xi, \kappa_\xi)\}.$$

This is clearly a projectable symmetry generator with projections

$$\vec{p}\restriction^* E_e \stackrel{\text{def}}{=} \langle \ldots, p_{\zeta_1} \cap (\kappa'_{\zeta_1} \times \alpha_1), \varnothing, \ldots, p_{\zeta_i} \cap (\kappa'_{\zeta_i} \times \alpha_i), \varnothing, \ldots \rangle$$

Let G be a $\mathbb{P}$-generic filter and take the symmetric model $V(G) \stackrel{\text{def}}{=} V(G)^{\mathcal{F}_I}$. Similarly to Lemma 1.30 we can show that the approximation lemma holds for this model. As in the previous section, the approximation lemma enables us to prove all the nice properties of $V(G)$. Note that $\mathbb{P}$ has the κ_0-cc., so all cardinals $\leq \kappa_0$ are preserved.

LEMMA 1.35. *In $V(G)$, for every $\xi \in (0, \rho)$, $(\kappa'_\xi)^+ = \kappa_\xi$, $\mathsf{cf}(\kappa'_\xi) = (\mathsf{cf}(\kappa'_\xi))^V$, and $\mathsf{cf}(\kappa_\xi) = (\mathsf{cf}(\kappa_\xi))^V$.*

PROOF. First we will show that for every $\xi \in (0, \rho)$, the interval $(\kappa'_\xi, \kappa_\xi)$ has collapsed to κ'_ξ, then that κ_ξ has not collapsed, and lastly we will show that the cofinalities of κ'_ξ and of κ_ξ are preserved. Fix $\xi \in (0, \rho)$ and let $\alpha \in (\kappa'_\xi, \kappa_\xi)$ be arbitrary. The name

$$\tau \stackrel{\text{def}}{=} \{(\check{\vec{p}}, \vec{p}) \,;\, \vec{p} \in E_\alpha\}$$

is a hereditarily symmetric $\mathbb{P}$-name and $\bigcup(\tau^G(\xi))$ is a surjection from a subset of κ'_ξ onto α. So it remains to show that κ_ξ has not collapsed. Assume towards a contradiction that in $V(G)$ there is a $\beta < \kappa_\xi$ and a bijective function $f : \beta \to \kappa_\xi$. Such f can be seen as a set of ordinals of size κ_ξ, therefore by the approximation lemma for $V(G)$ there is an $E_e \in I$ with $e = \{\alpha_1, \ldots, \alpha_m\}$ such that

$$f \in V[G \cap E_e].$$

So f is forced by

$$E'_e = \prod_{i=1,\dots,m} \mathsf{Fn}(\kappa'_{\zeta_i}, \alpha_i, \kappa'_{\zeta_i}),$$

because this is isomorphic to E_e. We will identify E_e with E'_e. Let j be the greatest such that $\kappa_\xi > \alpha_j$. We have that

$$E_e = \prod_{i=1,\dots,j} \mathsf{Fn}(\kappa'_{\zeta_i}, \alpha_i, \kappa'_{\zeta_i}) \times \prod_{i=j+1,\dots,m} \mathsf{Fn}(\kappa'_{\zeta_i}, \alpha_i, \kappa'_{\zeta_i})$$

The image below shows a closeup of a 'worst case scenario' situation.

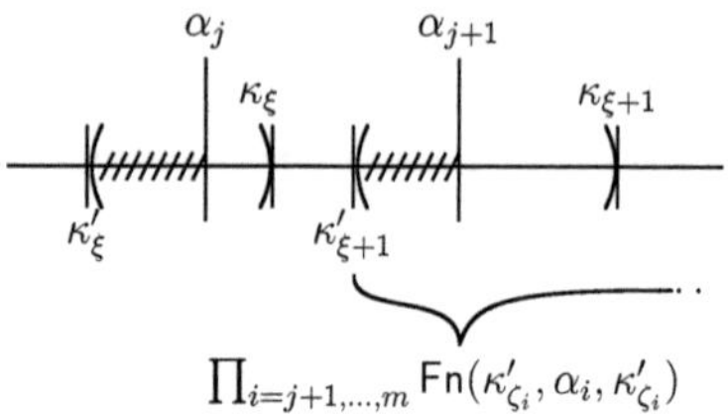

Since for every $i = j+1, \dots, m$, κ'_{ζ_i} is regular, $\mathsf{Fn}(\kappa'_{\zeta_i}, \alpha_i, \kappa_{\zeta_i})$ is κ'_{ζ_i}-closed. So by Lemma 1.11, the finite product

$$\prod_{i=j+1,\dots,m} \mathsf{Fn}(\kappa'_{\zeta_i}, \alpha_i, \kappa'_{\zeta_i})$$

is κ_ξ-closed. So by the same lemma, f must be forced by just the partial order

$$\prod_{i=1,\dots,j} \mathsf{Fn}(\kappa'_{\zeta_i}, \alpha_i, \kappa'_{\zeta_i}).$$

But for each $i = 1, \dots, j$, $\mathsf{Fn}(\kappa'_{\zeta_i}, \alpha_i, \kappa'_{\zeta_i})$ has the $(|\alpha_i|^{<\kappa'_{\zeta_i}})^+$-cc and since κ_ξ is inaccessible, the κ_ξ-cc. So by Lemma 1.11 again, the partial order

$$\prod_{i=1,\dots,j} \mathsf{Fn}(\kappa'_{\zeta_i}, \alpha_i, \kappa'_{\zeta_i})$$

has the κ_ξ-cc, therefore it cannot add such an f either. Contradiction.

With a similar analysis we can show that for each $\xi \in (0, \rho)$, the cofinalities of κ_ξ and κ'_ξ are the same in $V(G)$ as they are in V. qed

With the same method of splitting the support E_e into the part below and the part above a κ_ξ, we can show that the pattern of cardinals in $V(G)$ is as we intended.

LEMMA 1.36. *In $V(G)$, for each $\xi \in (0, \rho)$, κ_ξ is a measurable cardinal (with a normal measure). Therefore in $V(G)$ the cardinals in the interval $[\kappa_0, \bigcup_{\xi<\rho} \kappa_\xi]$ are alternating regular and measurable cardinals, with the exception of their singular limits.*

PROOF. Let $\xi \in (0, \rho)$ be arbitrary and let U be a κ_ξ-complete ultrafilter on κ_ξ. In $V(G)$ define the set

$$W \stackrel{\text{def}}{=} \{x \subseteq \kappa_\xi \,;\, \exists y \in U (y \subseteq x)\}.$$

This is clearly a filter in $V(G)$ so it remains to show that it is an ultrafilter and that it is κ_ξ-complete. So let $X \subseteq \kappa_\xi$ be arbitrary in $V(G)$ and let $\dot{X} \in \mathsf{HS}$ be a name for X, supported by $E_e \in I$. By the approximation lemma, we have that

$$X \in V[G \cap E_e].$$

So X is added by the forcing E_e, which as in the previous proof can be split to the part above κ_ξ and the part below κ_ξ. The part above does not add bounded subsets to $\kappa'_{\xi+1}$ which is strictly larger than κ_ξ so X must have been added by the part of E_e below κ_ξ. But that is small forcing with respect to κ_ξ so we get that $X \in W$ or $\kappa_\xi \setminus X \in W$. Similarly we show that it is κ_ξ -complete and that if U is normal then W is normal as well (see also proof of Lemma 1.33 and the comments afterwards). qed

This model does not satisfy AC. Just as in Lemma 1.32 we have that in $V(G)$, for every $\xi \in (0, \rho)$, the powerset of κ'_ξ is a κ_ξ-long union of sets with size $\leq \kappa'_\xi$. Moreover we have the following.

LEMMA 1.37. *In $V(G)$, $\mathsf{AC}_{\kappa_1}(\mathcal{P}(\kappa_0))$ fails, i.e., in $V(G)$ there is a non empty family of κ_1-sized subsets of $\mathcal{P}(\kappa_0)$ without a choice function.*

PROOF. Since κ_1 is the successor of κ_0 in $V(G)$, by Lemma 0.3 there is a surjection $f : \mathcal{P}(\kappa_0) \to \kappa_1$. Assume towards a contradiction that $\mathsf{AC}_{\kappa_1}(\mathcal{P}(\kappa_0))$ holds. Since $\kappa_1 = \kappa_0^+$ in $V(G)$, by Lemma 0.2 there exists an injective function $f' : \kappa_1 \to \mathcal{P}(\kappa_0)$. Since κ_1 is measurable in $V(G)$, let W_1 be a κ_1-complete ultrafilter over κ_1. Define a function $h : \kappa_1 \to W_1$ by

$$h(\alpha) \stackrel{\text{def}}{=} \begin{cases} \{\xi < \kappa_1 \,;\, \alpha \in f'(\xi)\} & \text{if this is in } W_1 \\ \{\xi < \kappa_1 \,;\, \alpha \notin f'(\xi)\} & \text{otherwise} \end{cases}$$

Since W_1 is κ_1-complete, $X \stackrel{\text{def}}{=} \bigcup_{\alpha<\kappa_1} h(\alpha) \in W_1$. But X has at most one element, contradiction. qed

By using such a restricted symmetry generator we make a lot of arguments easier because this model is approximated by finite products of collapsing functions (see for example Lemma 3.53). The downside to this method is that we cannot easily disprove choice statements. For example we conjecture that DC fails in this model (even when $\kappa_0 > \omega_1$), but such a proof is not obvious.

To get a model in which we have alternating measurables and DC holds we could modify this construction as follows.

For every $e = \langle e_\xi \,;\, \xi \in (0, \rho) \rangle \in \prod_{\xi \in (0,\rho)} (\kappa'_\xi, \kappa_\xi)$ define

$$E'_e \stackrel{\text{def}}{=} \{\langle p_\xi \cap (\kappa'_\xi \times e_\xi) \,;\, \xi \in (0, \rho) \rangle \,;\, \vec{p} = \langle p_\xi \,;\, \xi \in (0, \rho) \in \mathbb{P} \rangle\}, \text{ and}$$

$$I' \stackrel{\text{def}}{=} \{E'_e \,;\, e \in \prod_{\xi \in (0,\rho)} (\kappa'_\xi, \kappa_\xi)\}.$$

Then if we take $V(G)' \stackrel{\text{def}}{=} V(G)^{\mathcal{F}_{I'}}$ we end up with a model similar to the model constructed for [**AK06**, Theorem 5]. With the same proofs we can show Lemma 1.35 and Lemma 1.36 for $V(G)'$. With a similar proof as [**Apt83a**, Lemma 1.4] we can show the following.

LEMMA 1.38. *In* $V(G)'$, *for every* $\alpha < \kappa_0$, DC_α *holds, i.e., for every set* $X \in V(G)'$ *with a binary relation* $R \in V(G)'$, *if for every* $\gamma < \alpha$ *and for every R-sequence* $\langle x_\beta \,;\, \beta < \gamma\rangle$ *of elements of* X *there is an* $x \in X$ *such that* $\forall \beta < \gamma (x_\beta R x)$ *holds, then there exists an* R *sequence of length* α.

Another similar construction to these can be found in [**AJL**, Theorem 32] where for three successive cardinals the pattern "measurable-regular-measurable" is symmetrically forced.

2

Patterns of singular cardinals of cofinality ω

In this chapter we will look at the consistency strength of several patterns of singular cardinals. It is a theorem of ZFC that all successor cardinals are regular. So in order to force several patterns of singular cardinals we have to drop once more the axiom of choice.

We will be looking at singular cardinals of cofinality ω because they are created via simple versions of the Prikry forcing (Prikry forcing, tree-Prikry forcing, strongly compact tree-Prikry forcing) which interfere very little with the surrounding cardinals. This is because these forcings do not add bounded subsets to the cardinal that has to become singular and that will enable us to use the tricks of splitting a product just as in Lemma 1.36.

In the first section we will take a look at the tools necessary for creating the symmetric models in this chapter. We will discuss versions of Prikry type forcings that are appropriate for applying automorphisms to, and will explain why these forcings work just like the usual Prikry type forcings found in the handbook of set theory [**Git10**].

In Section 2 we will construct arbitrary long sequences of alternating regular and singular cardinals of cofinality ω. This is the simplest pattern in the sense that we can construct this from just ZFC.

In Section 3 is inspired by a question of Benedikt Löwe. There we will construct a countable sequence of any desirable pattern of regular cardinals and singular cardinals of cofinality ω. In particular for any function $f : \omega \to 2$ in the ground model, we will construct a symmetric model in which $\aleph_{n+1}$ is singular of cofinality ω if $f(n) = 0$ and regular if $f(n) = 1$. Here will be our first new results connected with almost Ramsey cardinals and Rowbottom cardinals that carry Rowbottom filters (the last part in cooperation with Arthur Apter). For this symmetric model we will use an ω-sequence of strongly compact cardinals.

In Section 4 we will show how to construct longer countable sequences of successive singular cardinals, thus showing how to deal with limit points. We will do this in an attempt to make the next section more approachable.

In the last section we'll be looking at a modification of Gitik's model in *All uncountable cardinals can be singular* [**Git80**], that also inspired sections 3 and 4. There we will do a symmetric forcing construction that will result on a ρ-long sequence of successive singular cardinals, where ρ is any predetermined cardinal.

This section is very much connected to the question posed by Apter "Which cardinals can become simultaneously the first measurable and the first regular uncountable cardinal?". Apter had worked on this question in [**Apt96**] where he showed that $\aleph_2$ can be the first regular and the first measurable cardinal simultaneously. The methods of that paper could be extended to construct models in which $\aleph_{\omega+1}$ (or $\aleph_{\omega+2}$) is both the least measurable and least regular cardinal. That technique could not be extended to other cardinals, e.g. to $\aleph_3$.

Following the author's study and modification of Gitik's paper "All uncountable cardinals can be singular" [**Git80**], Apter noticed that this modification of Gitik's construction is a generalisation of the techniques he used in [**Apt85**], with Henle in [**AH91**], and with Magidor in [**?**], in order to collapse a measurable cardinal to become the successor of a singular cardinal. He therefore suggested to the author to "end" the construction with just a measurable cardinal and get in the resulting model that every cardinal below that measurable is singular. Thus the measurable will be the first regular and first measurable cardinal simultaneously. This model is constructed from a ρ-long sequence of strongly compact cardinals. With a simple modification of the ultrafilters used in Gitik's construction the author was able to make sure the intervals between the strongly compacts collapse. Moreover, with a less simple (technically) construction (see Theorem 2.37), the author was able to prove that, as expected, none of the former strongly compacts have collapsed in that model. Thus the height at which the measurable cardinal would be is determined exactly, i.e., we end up with a symmetric model in which $\aleph_{\rho+1}$ is the first measurable and first regular uncountable cardinal. This construction and proofs are in the last section of Chapter 2, it is joint work with Arthur Apter and Peter Koepke, and it is submitted for publication [**ADK**].

In the end of the section we'll explain how this result generalises to other large cardinal properties that, as measurability, are preserved after small symmetric forcing.

1. Prikry-type forcings for symmetric forcing

In this section we will present the forcings that we will use in the construction of this chapter and prove some basic facts that we will need later. Some of our constructions will use finite support products of tree-Prikry forcings. This forcing is described in [**Git10**], but we will describe it slightly modified so we can apply permutations to it and create symmetric models.

In this and the next section we will use strongly compact cardinals. We will use the fine ultrafilters that these cardinals give to induce certain measures. In particular, for every regular α we will use a surjection $h : \mathcal{P}_\kappa(\alpha) \to \alpha$ and a fine ultrafilter U over $\mathcal{P}_\kappa(\alpha)$ to induce the following κ-complete ultrafilter Φ_U over α,

(1) $$\Phi_{U,h} \stackrel{\text{def}}{=} \{X \subseteq \alpha\ ;\ h^{-1}``X \in U\}.$$

It's easy to check that if U is a fine ultrafilter over $\mathcal{P}_\kappa(\alpha)$ then Φ_U, as defined above, is a κ-complete ultrafilter over α and it is uniform, i.e., all its elements have size α.

For the next sections we will need to make a strongly compact cardinal κ singular while collapsing a particular interval directly above it. To do that in a way we can apply permutations later we use the following type of tree-Prikry forcing.

DEFINITION 2.1. Let κ be a measurable cardinal, let $\alpha \geq \kappa$ be a regular cardinal, and Φ a uniform κ-complete ultrafilter over α. A set $T \subseteq {}^{<\omega}\alpha$ is called an injective Φ-tree iff

(1) T consists of finite injective sequences of elements of α,
(2) T is a tree with respect to end extension "$\trianglelefteq$",
(3) T has a trunk, i.e., an element denoted by tr_T, that is maximal in T such that for every $t \in T$, $t \trianglelefteq \mathsf{tr}_T$ or $\mathsf{tr}_T \trianglelefteq t$, and
(4) for every $t \in T$ with $t \trianglerighteq \mathsf{tr}_T$, the set
$$\mathsf{Suc}_T(t) \stackrel{\text{def}}{=} \{\beta \in \alpha\ ;\ t^\frown\langle\beta\rangle \in T\}$$
is in the ultrafilter Φ.

We will call our forcing $\mathbb{P}^{\mathsf{t}}_\Phi$ injective tree-Prikry forcing with respect to the ultrafilter Φ. It consists of all injective Φ-trees, and it is ordered by direct inclusion, i.e., $T \leq S \stackrel{\text{def}}{\Longleftrightarrow} T \subseteq S$. If T is an injective Φ-tree and $t \in T$ is such that $\mathsf{tr}_T \trianglelefteq t$ then define
$$T \uparrow t \stackrel{\text{def}}{=} \{t' \in T\ ;\ t' \trianglelefteq t \text{ or } t \trianglelefteq t'\},$$
the least extension of T that has t as a trunk. Clearly, $T \uparrow t \leq T$.

In the definition above, uniformity is very important because it ensures the following lemma.

LEMMA 2.2. *The injective tree-Prikry forcing $\mathbb{P}^{\mathsf{t}}_\Phi$ does not add bounded subsets to κ and has the α^+-cc.*

Similarly to the standard tree-Prikry forcing (see [**Git10**, Lemma 1.24 and Lemma 1.9]) we see that $\mathbb{P}^{\mathsf{t}}_\Phi$ adds an ω sequence cofinal in α.

LEMMA 2.3. *If G is a generic filter over $\mathbb{P}^{\mathsf{t}}_\Phi$, then in $V[G]$ the ordinal α has cofinality ω.*

PROOF. For every $\beta < \alpha$ define the set
$$D_\beta \stackrel{\text{def}}{=} \{T \in \mathbb{P}^{\mathsf{t}}_\Phi\ ;\ \exists n \in \mathsf{dom}(\mathsf{tr}_T)(\mathsf{tr}_T(n) \geq \beta)\}.$$
To see that this set is dense in $\mathbb{P}^{\mathsf{t}}_\Phi$, let $\beta < \alpha$ and let $S \in \mathbb{P}^{\mathsf{t}}_\Phi \setminus D_\beta$. For each $n \in \mathsf{dom}(\mathsf{tr}_S)$ we have that $\mathsf{tr}_S(n) \neq \beta$. The set $\mathsf{Suc}_S(\mathsf{tr}_S) = \{\gamma < \alpha\ ;\ \mathsf{tr}_S{}^\frown\langle\gamma\rangle \in S\}$ is in the ultrafilter Φ_α, which is uniform. So there must be some $\gamma \geq \beta$ in $\mathsf{Suc}_S(\mathsf{tr}_S)$. Fix this γ. The condition $S \uparrow (\mathsf{tr}_S{}^\frown\langle\gamma\rangle)$ is in D_β and it is stronger than S. So D_β is dense in $\mathbb{P}^{\mathsf{t}}_\Phi$. qed

To prove that the cardinals between κ and α^+ collapse we use an ultrafilter Φ over α that is induced by a fine ultrafilter over $\mathcal{P}_\kappa(\alpha)$. To get a forcing that is isomorphic to the injective tree-Prikry forcing, we will use an inaccessible α. This is because we know that if α is an inaccessible cardinal and $\kappa \leq \alpha$ is a cardinal, then there is a bijection $h_\alpha : \mathcal{P}_\kappa(\alpha) \to \alpha$. Such a bijection will create an isomorphism between injective tree-Prikry forcing and the defined below injective strongly compact tree-Prikry forcing. The necessity of the inaccessible is not a problem when we want to collapse an interval (κ, κ') where κ' is a limit of inaccessible cardinals. That is because strongly compact cardinals are such limits.

DEFINITION 2.4. Let α be inaccessible, $\kappa \leq \alpha$ be α-strongly compact, and U a κ-complete fine ultrafilter over $\mathcal{P}_\kappa(\alpha)$. The injective strongly compact tree-Prikry forcing $\mathbb{P}^{\mathsf{st}}_U$ with respect to U consists of all injective U-trees, i.e., of all $T \subseteq {}^{<\omega}\mathcal{P}_\kappa(\alpha)$ such that

(1) T consists of finite injective sequences of elements of $\mathcal{P}_\kappa(\alpha)$,
(2) T is a tree with respect to end extension "$\trianglelefteq$",
(3) T has a trunk, and
(4) for every $t \in T$ with $t \trianglerighteq \mathsf{tr}_T$, the set

$$\mathsf{Suc}_T(t) \stackrel{\text{def}}{=} \{b \in \mathcal{P}_\kappa(\alpha)\ ;\ t^\frown\langle b\rangle \in T\}$$

is in the ultrafilter U.

The ordering in this forcing is also defined as $T \leq S \stackrel{\text{def}}{\Longleftrightarrow} T \subseteq S$.

It is not hard to prove that the intended isomorphism between an injective tree-Prikry forcing and an injective strongly compact tree-Prikry forcing exists.

PROPOSITION 2.5. *Let α be an inaccessible cardinal, $\kappa \leq \alpha$ a strongly compact cardinal, U a fine ultrafilter over $\mathcal{P}_\kappa(\alpha)$, $h : \mathcal{P}_\kappa(\alpha) \to \alpha$ a bijection, and $\Phi_{U,h}$ the κ-complete ultrafilter over α that is induced by U and h as shown in (1). Then, the injective tree-Prikry forcing $\mathbb{P}^{\mathsf{t}}_{\Phi_{U,h}}$ with respect to $\Phi_{U,h}$ is isomorphic to the injective strongly compact tree-Prikry forcing $\mathbb{P}^{\mathsf{st}}_U$.*

Similarly to the standard strongly compact tree-Prikry forcing (see [**Git10**, Lemma 1.50]) we can show the following.

PROPOSITION 2.6. *If α is an inaccessible cardinal, $\kappa \leq \alpha$ is α-strongly compact, and U is a fine ultrafilter over $\mathcal{P}_\kappa(\alpha)$ then the forcing $\mathbb{P}^{\mathsf{st}}_U$ collapses α to κ by making it a countable union of κ-sized sets.*

Using the two propositions above we get the following.

COROLLARY 2.7. *With the notation of Proposition 2.5, $\mathbb{P}^{\mathsf{t}}_{\Phi_{U,h}}$ preserves all cardinals $\leq \kappa$ and above α and collapses α to κ while making κ singular.*

It is not obvious whether these injective versions of tree-Prikry forcing and strongly compact tree-Prikry forcing completely or densely embed into the standard ones, or the other way around.

Finally, let us see how permutations of ordinals will be used to get automorphisms of the injective tree-Prikry forcing. For the rest of this section fix κ a strongly compact cardinal, an ordinal $\alpha \geq \kappa$, a uniform κ-complete ultrafilter Φ over α, and let $\mathbb{P}^{\mathsf{t}}_{\Phi}$ be the injective tree-Prikry forcing with respect to Φ. Let $\mathcal{G}_\alpha$ be the group of those automorphisms a of α that move only finitely many ordinals. Call $\mathsf{supp}(a)$ the finite set of ordinals in α that a moves.

Definition 2.8. For $a \in \mathcal{G}_\alpha$, $T \in \mathbb{P}^{\mathsf{t}}_{\Phi}$, and $t \in T$, define

$$a``t \stackrel{\text{def}}{=} \{(n, a(\beta)) \,;\, (n,\beta) \in t\}, \text{ and}$$
$$a```T \stackrel{\text{def}}{=} \{a``t \,;\, t \in T\}.$$

Proposition 2.9. *The map $T \mapsto a```T$ is an automorphism of $\mathbb{P}^{\mathsf{t}}_{\Phi}$.*

Proof. The set $a```T$ is a tree with respect to $\trianglelefteq$, it has a trunk, and $\mathsf{tr}_{a```T} = a``\mathsf{tr}_T$. If $t \in a```T$ is such that $t \trianglerighteq \mathsf{tr}_{a```T}$ then there is a $t' \in T$ such that $t = a``t'$. We have that

$$\begin{aligned}\{\beta < \alpha \,;\, t^\frown\langle\beta\rangle \in a```T\} &= \{\beta < \alpha \,;\, (a``t)^\frown\langle\beta\rangle \in a```T\} \\ &\supseteq \{\beta < \alpha \,;\, t'^\frown\langle\beta\rangle \in T\} \setminus \mathsf{supp}(a) \\ &\in \Phi.\end{aligned}$$

So $a```T \in \mathbb{P}^{\mathsf{t}}_{\Phi}$. That a is indeed a morphism is easy to check. qed

Now we have all the necessary tools to proceed to our constructions.

2. Alternating regular and singular cardinals

In this section we prove the following theorem.

Theorem 2.10. *If V is a model of* ZFC*, κ_0 is a regular cardinal in V and ρ is an ordinal in V, then there is a model of* ZF *with a sequence of successive alternating singular and regular cardinals that starts at κ_0 and that contains ρ-many singular cardinals.*

The model we construct is the same as the one in Section 4 of Chapter 1. The result would be the same if we did a Feferman-Lévy type construction, i.e., instead of collapsing the entire interval $(\kappa'_\xi, \kappa_\xi)$ (where here κ_ξ is a singular cardinal), we collapse to κ'_ξ just a sequence that is cofinal to κ_ξ.

We start from a model V of ZFC. In the model of Section 4 of Chapter 1 we had inaccessible cardinals κ_ξ for $\xi \in (0, \rho)$. Here the κ_ξ are just singular cardinals, so we assume GCH in our ground model, in order to have the appropriate chain conditions for the parts of our forcing. Let ρ be an ordinal, and $\langle \kappa_\xi \,;\, \xi \in \rho\rangle$ be an increasing sequence of cardinals such that κ_0 is regular and for each $0 < \xi < \rho$, κ_ξ is a singular cardinal. We will collapse every $\kappa_{\xi+1}$ to become the successor of $(\kappa_\xi)^+$, thus constructing an alternating sequence of consecutive singular and regular cardinals.

For every $\xi \in (0, \rho)$ define κ'_ξ just as we did in page 38 and the same for $\mathbb{P}$, $\mathcal{G}$, E_e, I, and $V(G) = V(G)^{\mathcal{F}_I}$. Note that the last case of κ'_ξ in that page never occurs here.

The approximation lemma holds for the same reasons as in Section 4 and since we have the GCH, Lemma 1.35 and Lemma 1.36 hold here as well, so the cardinal pattern is as we wanted. For the same reasons, Lemma 1.37 holds as well, so the powerset of κ_0 is non wellorderable and in particular $\mathsf{AC}_{\kappa_1}(\mathcal{P}(\kappa_0))$ fails. With singular cardinals being successor cardinals, the situation is actually even stranger here.

LEMMA 2.11. *In $V(G)$, for every $\xi \in [0,\rho)$, the powerset of κ'_ξ is a κ'_ξ-sized union of κ'_ξ-sized sets, i.e., $\mathsf{US}(\kappa'_\xi)$ holds.*

PROOF. Fix an arbitrary $\xi \in (0,\rho)$. For every $(x \subseteq \kappa'_\xi)^{V(G)}$, use AC in the outer model to fix a name $\dot{x} \in \mathsf{HS}$ for x and a support E_{e_x} for $\dot{x}$. Since for every $e \in \prod^{\mathsf{fin}}_{\zeta\in[\xi,\rho)}(\kappa'_\zeta,\kappa_\zeta)$, E_e does not add new subsets to κ_ξ, we may assume that $e \in \prod^{\mathsf{fin}}_{\zeta\in(0,\xi)}(\kappa'_\zeta,\kappa_\zeta)$.

For each $e \in \prod^{\mathsf{fin}}_{\zeta\in(0,\xi)}(\kappa'_\zeta,\kappa_\zeta)$ define

$$C_e \stackrel{\text{def}}{=} \{x \in (\mathcal{P}(\kappa_\xi))^{V(G)} \,;\, E_e \text{ supports } \dot{x}\}.$$

We have that in $V(G)$

$$\mathcal{P}(\kappa_\xi) = \bigcup_{e\in\prod^{\mathsf{fin}}_{\zeta\in(0,\xi)}(\kappa'_\zeta,\kappa_\zeta)} C_e.$$

We want to prove that in $V(G)$ each such C_e has size κ'_ξ. Work in V. For every $e \in \prod^{\mathsf{fin}}_{\zeta\in(0,\xi)}(\kappa'_\zeta,\kappa_\zeta)$ and every $x \in C_e$ define a name

$$\ddot{x} \stackrel{\text{def}}{=} \{(\check{\beta},\vec{p}\!\upharpoonright^{*}\!E_e) \,;\, \vec{p} \Vdash \check{\beta} \in \dot{x}\}.$$

The approximation lemma gives us that $(\dot{x})_G = (\ddot{x})_G = x$. For each $e \in \prod^{\mathsf{fin}}_{\zeta\in(0,\xi)}(\kappa'_\zeta,\kappa_\zeta)$ define

$$C'_e \stackrel{\text{def}}{=} \{\ddot{x} \,;\, x \in C_e\},$$

and note that the function that sends each x to $\ddot{x}$ is an injection from C_e into C'_e. Thus it suffices to show that $V(G) \models$ "C'_e has cardinality κ'_ξ". We have that in V, $C'_e \subseteq \mathcal{P}^{(k)}(\kappa'_\xi)$ for some finite k. Since GCH holds in V we have that C'_e is of size $\kappa_\xi^{'(+k)}$ in V, and because κ_ξ is a limit cardinal, we have that $\kappa_\xi^{'(+k)}$ has size κ'_ξ in $V(G)$. So in $V(G)$, the powerset of κ'_ξ is a κ_ξ-union of κ'_ξ sized sets.

Since for each $\xi \in (0,\rho)$ we have that κ_ξ is singular, we have that the powerset of every κ'_ξ can be written as a $\leq \kappa'_\xi$-sized union of $\leq \kappa'_\xi$-sized sets. qed

By Lemma 0.5, $\mathsf{US}(\kappa'_\xi)$ implies that κ_ξ is singular, (which we already knew of course) and that every wellorderable subset of κ'_ξ has cardinality κ'_ξ. By Lemma 0.4 we have that for every $\xi \in (0,\rho)$ there are no surjections from $\mathcal{P}(\kappa_\xi)$ onto $(\kappa'_{\xi+1})^{\kappa_\xi})$. By the diagram of page 16 $\mathsf{AC}_{\kappa_\xi}(\mathcal{P}(\kappa_\xi))$ fails for every $\xi \in [0,\rho)$, thus also $\mathsf{AC}_{\kappa_0}(\mathcal{P}(\kappa_0))$ fails and $\mathcal{P}(\kappa_0)$ is non-wellorderable.

3. An arbitrary ω-long pattern of singular and regular cardinals

According to **[Sch99]** having a model of ZF with two consecutive singular cardinals implies the existence of an inner model with a Woodin cardinal. So we cannot symmetrically force patterns of singular cardinals that include consecutive singular cardinals, just from a model of ZFC. For such a construction we need large cardinals.

In this section we construct a countable sequence of cardinals, in any desirable pattern of regular cardinals and singular cardinals of cofinality ω from a sequence of strongly compact cardinals. In particular we prove the following theorem.

THEOREM 2.12. *Assume that V is a model of ZFC with an increasing sequence $\langle \kappa_n \,;\, 0 < n < \omega \rangle$ of strongly compact cardinals, which sequence has limit η. For any function $f : \omega \to 2$ of the ground model, there is a model of ZF in which $\aleph_{n+1}$ is regular if $f(n) = 1$ and singular if $f(n) = 0$.*

This construction, and the constructions in the next sections are inspired by Moti Gitik's *All uncountable cardinals can be singular* [**Git80**].

Let V be a model of ZFC in which there is an increasing sequence

$$\langle \kappa_n \,;\, 0 < n < \omega \rangle$$

of strongly compact cardinals, with limit η, and let $f : \omega \to 2$ be an arbitrary function which we will use to create our pattern. We will create a symmetric model in which $\aleph_{n+1}$ will be regular if $f(n) = 1$ and singular of cofinality ω if $f(n) = 0$. We take an ω-long sequence of strongly compacts for ease of notation but in fact we only need as many as the cardinality of the set $\{n \in \omega \,;\, f(n) = f(n+1) = 1\}$, i.e., as many as the amount of successive singular cardinals we want to have.

Define the set Reg^η to be the set of regular cardinals in the interval (ω, η). To get an idea of what is to come, let us describe the first steps we will take. First we must collapse κ_1 to become the successor of ω by adding surjections from ω to α, for every $\alpha \in (\omega, \kappa_1)$. If $f(0) = 0$ then we must make κ_1 singular of cofinality ω using an injective tree-Prikry forcing for every regular cardinal in (κ_1, κ_2). This will also make sure that κ_2 becomes the successor of κ_1. If $f(0) = 1$ then we just symmetrically collapse κ_2 to become the successor of κ_1, and so on.

So in general, for every $\alpha \in \mathsf{Reg}^\eta$, define the following forcings.

- If for some $n < \omega$, $\alpha \in [\kappa_{n+1}, \kappa_{n+2})$ and $f(n) = 0$ then let H_α be a fine ultrafilter over $\mathcal{P}_{\kappa_{n+1}}(\alpha)$, and $h_\alpha : \mathcal{P}_{\kappa_{n+1}}(\alpha) \to \alpha$ be a surjection. If α is inaccessible then let h_α be moreover a bijection. Note that the set

$$\Phi_\alpha \stackrel{\text{def}}{=} \Phi_{H_\alpha, h_\alpha} = \{X \subseteq \alpha \,;\, h_\alpha^{-1}``X \in H_\alpha\}$$

  is a κ_{n+1}-complete ultrafilter over α. So let $\mathbb{P}_\alpha \stackrel{\text{def}}{=} \mathbb{P}^{\mathrm{t}}_{\Phi_\alpha}$ be injective tree-Prikry forcing, as defined in Definition 2.1, with respect to the ultrafilter Φ_α.
- If for some $n < \omega$, $\alpha \in [\kappa_{n+1}, \kappa_{n+2})$ and $f(n) = 1$ then let

$$\mathbb{P}_\alpha \stackrel{\text{def}}{=} \{p : \kappa_{n+1} \rightharpoonup \alpha \,;\, |p| < \kappa_{n+1}\}.$$

- If $\alpha \in (\omega, \kappa_1)$ then let

$$\mathbb{P}_\alpha \stackrel{\text{def}}{=} \{p : \omega \rightharpoonup \alpha \,;\, |p| < \omega\}.$$

For ease of notation, define the following sets.

$$\mathsf{Pr} \stackrel{\text{def}}{=} \{\alpha \in \mathsf{Reg}^\eta \,;\, \exists n \in \omega(\alpha \in [\kappa_{n+1}, \kappa_{n+2}) \text{ and } f(n) = 0)\}, \text{ and}$$
$$\mathsf{Co} \stackrel{\text{def}}{=} \{\alpha \in \mathsf{Reg}^\eta \,;\, \exists n \in \omega(\alpha \in [\kappa_{n+1}, \kappa_{n+2}) \text{ and } f(n) = 1) \text{ or } \alpha \in (\omega, \kappa_1)\}$$

We will force with the finite support product of these $\mathbb{P}_\alpha$, i.e., with

$$\mathbb{P} \stackrel{\text{def}}{=} \prod_{\alpha \in \mathsf{Reg}^\eta}^{\text{fin}} \mathbb{P}_\alpha,$$

which is partially ordered by

$$\vec{T} = \langle T_\alpha \,;\, \alpha \in \mathsf{dom}(\vec{T})\rangle \leq \vec{S} = \langle S_\alpha \,;\, \alpha \in \mathsf{dom}(\vec{S})\rangle$$
$$\stackrel{\text{def}}{\iff}$$
$$\forall \alpha \in \mathsf{Pr}(T_\alpha \subseteq S_\alpha) \text{ and } \forall \alpha \in \mathsf{Co}(T_\alpha \supseteq S_\alpha).$$

Vaguely speaking, for a condition $\vec{T} = \langle T_\alpha \,;\, \alpha \in \mathsf{dom}(\vec{T})\rangle$ we will permute each part T_α of the condition separately in a way that the result stays in $\mathbb{P}$. Then we will fix finite subsets of Reg^η and so get a symmetric model. All should be very controlled by being finite.

Formally, for each $\alpha \in \mathsf{Reg}^\eta$, let $\mathcal{G}_\alpha$ be the group of permutations of α that move only finitely many elements of α. Let $\mathcal{G}$ be the finite support product of all these $\mathcal{G}_\alpha$, i.e.,

$$\mathcal{G} \stackrel{\text{def}}{=} \prod_{\alpha \in \mathsf{Reg}^\eta}^{\text{fin}} \mathcal{G}_\alpha.$$

We denote an element of $\mathcal{G}$ by $\vec{a} = \langle a_\alpha \,;\, \alpha \in \mathsf{Reg}^\eta\rangle$, where only for finitely many of the $\alpha \in \mathsf{Reg}^\eta$ a_α is not the identity.

We extend $\mathcal{G}$ to act on $\mathbb{P}$ as follows. For $\vec{T} \in \mathbb{P}$ and $\vec{a} = \langle a_\alpha \,;\, \alpha \in \mathsf{Reg}^\eta\rangle \in \mathcal{G}$, and $\alpha \in \mathsf{dom}(\vec{T})$, let

$$\vec{a}(\vec{T}) \stackrel{\text{def}}{=} \begin{cases} a_\alpha ``` T_\alpha & \text{if } \alpha \in \mathsf{Pr} \\ a_\alpha `` T_\alpha & \text{if } \alpha \in \mathsf{Co} \end{cases}$$

where $a_\alpha ``` T_\alpha$ is as defined in Definition 2.8.

Note that as in Proposition 2.9 we get that $\vec{a}(\vec{T})$ is $\mathbb{P}$, the extended map $\vec{a}$ is an automorphism of $\mathbb{P}$, and $\mathcal{G}$ is now seen as an automorphism group of $\mathbb{P}$. We proceed to the definition of the symmetry generator we will use.

DEFINITION 2.13. For each finite $e \subseteq \mathsf{Reg}^\eta$ define

$$E_e \stackrel{\text{def}}{=} \{\vec{T} \in \mathbb{P} \,;\, \mathsf{dom}(\vec{T}) \subseteq e\}, \text{ and}$$
$$I \stackrel{\text{def}}{=} \{E_e \,;\, e \subseteq \mathsf{Reg}^\eta \text{ is finite}\}.$$

This I is a projectable symmetry generator, with projections

$$\vec{T}\restriction^* E_e = \langle T_\alpha \in \vec{T} \,;\, \alpha \in e \cap \mathsf{dom}(\vec{T})\rangle.$$

Let G be a $\mathbb{P}$-generic filter and take the symmetric model

$$V(G) \stackrel{\text{def}}{=} V(G)^{\mathcal{F}_{\mathcal{I}}}.$$

The approximation lemma holds for this symmetric model, but before we go on to show that $\mathbb{P}$ is $\mathcal{G}, I$-homogeneous we need the following observation.

PROPOSITION 2.14. *Assume that $\vec{T}$, $\vec{S}$ are such that $\mathsf{dom}(\vec{T}) = \mathsf{dom}(\vec{S})$, that for each $\alpha \in \mathsf{dom}(\vec{T}) \cap \mathsf{Pr}$ we have $\mathsf{tr}_{T_\alpha} = \mathsf{tr}_{S_\alpha}$, and that for each $\alpha \in \mathsf{dom}(T) \cap \mathsf{Co}$ we have $T_\alpha = S_\alpha$. Then the sequence $\langle T_\alpha \cap S_\alpha \,;\, \alpha \in \mathsf{dom}(\vec{T})\rangle$ is a condition in $\mathbb{P}$ and it is stronger than both $\vec{T}$, $\vec{S}$.*

In fact, this holds for any $< \kappa_1$-long sequence of conditions in $\mathbb{P}$ with the same trunks.

This holds because of the $< \kappa_1$-completeness of each Φ_α for each $\alpha \in \mathsf{Reg}^\eta \setminus \kappa$.

LEMMA 2.15. *For every $E_e \in I$, every $\vec{T} \in \mathbb{P}$, and every $\vec{S} \in \mathbb{P}$ such that $\vec{S} \leq \vec{T}\restriction^* E_e$, there is an automorphism $a \in \mathsf{fix}E_e$ such that $a(\vec{T}) \parallel \vec{S}$. Therefore $\mathbb{P}$ is $\mathcal{G}, I)$-homogeneous and the approximation lemma holds for $V(G)$.*

PROOF. For every $\alpha \in \mathsf{Co} \cap ((\mathsf{dom}(\vec{T}) \cap \mathsf{dom}(\vec{S})) \setminus e)$ we can find a permutation a_α of α such that for every $\xi \in \mathsf{dom}(T_\alpha) \cap \mathsf{dom}(S_\alpha)$ it sends $T_\alpha(\xi)$ to $S_\alpha(\xi)$.

For every $\alpha \in \mathsf{Pr} \cap ((\mathsf{dom}(\vec{T}) \cap \mathsf{dom}(\vec{S})) \setminus e)$ we can similarly find a permutation a_α of α such that for every $\xi \in \mathsf{dom}(\mathsf{tr}_{T_\alpha}) \cap \mathsf{dom}(\mathsf{tr}_{S_\alpha})$ it sends $\mathsf{tr}_{T_\alpha}(\xi)$ to $\mathsf{tr}_{S_\alpha}(\xi)$. In this case the image $\{a_\alpha``t \,;\, t \in T_\alpha\}$ is in $\mathbb{P}_\alpha$ because a_α moves only finitely many elements of α.

For any other $\alpha \in \mathsf{Reg}^\eta$, define a_α to be the identity.

It is not hard to see that $\vec{a} \stackrel{\text{def}}{=} \langle a_\alpha \,;\, \alpha \in \mathsf{Reg}^\eta\rangle$ is the automorphism we were looking for. qed

From this we get immediately that each $\kappa_n \in \mathsf{Pr}$ is singular in $V(G)$.

PROPOSITION 2.16. *In $V(G)$, all ordinals in Pr have cofinality ω.*

PROOF. Let $\alpha \in \mathsf{Pr}$. By Lemma 2.3 there is a $\mathbb{P}_\alpha$-name for an ω-sequence that is cofinal in α. Such a name is in HS and it is supported by $E_{\{\alpha\}}$. qed

Now we have to show that only the κ_n remain cardinals in the interval (ω, η). First, let us show that everything else has collapsed.

THEOREM 2.17. *In $V(G)$, for every $\alpha \in \mathsf{Pr}$, if $0 < n < \omega$ is such that $\alpha \in (\kappa_n, \kappa_{n+1})$ then α has collapsed to κ_n. Moreover, for every $\alpha \in \mathsf{Co}$, if $0 < n < \omega$ is such that $\alpha \in (\kappa_n, \kappa_{n+1})$ then α has collapsed to κ_n and if $\alpha \in (\omega, \kappa_1)$ then it is countable.*

PROOF. Let $\alpha \in \mathsf{Pr}$, and let $0 < n < \omega$ be such that $\alpha \in (\kappa_n, \kappa_{n+1})$. By Propositions 2.5 and 2.6 we get that α has collapsed to κ_n by a function f with a name $\dot{f} \in \mathsf{HS}$ supported by $E_{\{\alpha\}}$.

To show that if $\alpha \in \mathsf{Co}$ then α collapses to either the largest κ_n below it or to ω, again just check what $E_{\{\alpha\}}$ does. The union of an $E_{\{\alpha\}}$-generic filter is a collapsing surjection, so there is an $E_{\{\alpha\}}$-name for such a surjection. Any $E_{\{\alpha\}}$-name is a $\mathbb{P}$-name in HS. qed

It remains to show that for every $0 < n < \omega$, κ_n is still a cardinal in $V(G)$. For this we will use the approximation lemma and we will have to prove a Prikry-like lemma for the Prikry-like parts of our forcing. In particular we need the following.

LEMMA 2.18. *Let $c \subseteq \mathsf{Pr}$ be finite. Let $\tau_1, \ldots, \tau_k$ be E_c-names and φ a formula of set theory with k-many free variables. Then for every forcing condition $\vec{T} \in E_c$ there is a stronger condition $\vec{S} \in E_c$ such that $\mathsf{dom}(\vec{S}) = \mathsf{dom}(\vec{T})$, for each $\alpha \in \mathsf{dom}(\vec{S})$ we have that $\mathsf{tr}_{S_\alpha} = \mathsf{tr}_{T_\alpha}$, and $\vec{S}$ decides $\varphi(\tau_1, \ldots, \tau_k)$.*

PROOF. Let $\vec{T} \in E_c$ and $\{\alpha_0, \ldots, \alpha_n\} \stackrel{\text{def}}{=} \mathsf{dom}(\vec{T}) = c$. We enumerate the set

$$A \stackrel{\text{def}}{=} \prod_{i \in (n+1)} (\{\alpha_i\} \times (\omega \setminus \mathsf{dom}(\mathsf{tr}_{T_{\alpha_i}})))$$

by a function $x : \omega \to A$, such that for every $m \in \omega$ and every $\ell = 1, \ldots, n$, the set

$$(\{k \in \omega \,;\, \exists j \in [0, m]((k, \alpha_\ell) = x(j))\} \cup \mathsf{dom}(\mathsf{tr}_{T_{\alpha_j}}))$$

is a natural number. We need such an x so that each $x``j \cup \mathsf{dom}^2(t)$ is a potential $\mathsf{dom}^2(\mathsf{tr}_{\vec{S}})$ for some $\vec{S} \le \vec{T}$. This set A with the ordering x may be depicted as follows.

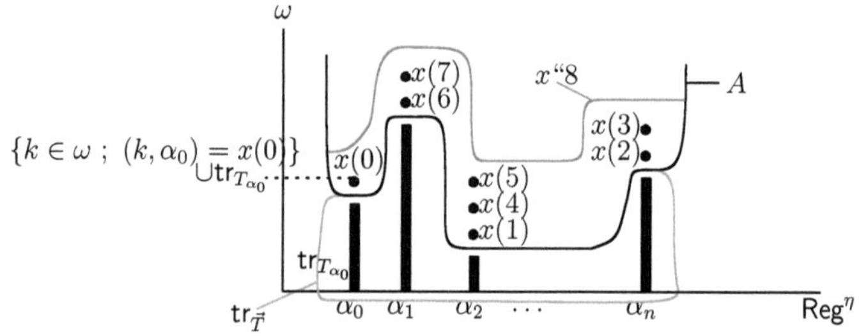

For all $\vec{s} = \langle s_{\alpha_0}, \ldots, s_{\alpha_n} \rangle$ define

$$\mathsf{dom}^2(\vec{s}) \stackrel{\text{def}}{=} \prod_{i \in (n+1)} (\{\alpha_i\} \times \mathsf{dom}(s_{\alpha_i})).$$

Let $\vec{t} \stackrel{\text{def}}{=} \langle \mathsf{tr}_{T_\alpha} \,;\, \alpha \in c \rangle$. We may also write $\mathsf{tr}_{\vec{T}} = \vec{t}$.

Possible trunks for extensions of $\vec{T}$ may have arbitrary lengths. To sort out the chaos, recursively define the set Σ as follows. A sequence $\vec{s} = \langle s_{\alpha_0}, \ldots, s_{\alpha_n} \rangle \in \Sigma \stackrel{\text{def}}{\iff}$

- $\vec{s} \in \prod_{i \in (n+1)} T_{\alpha_i}$,
- for some $i \in \omega$, $\mathsf{dom}^2(\vec{s}) \setminus \mathsf{dom}^2(\vec{t}) = x``i$
- if $i > 0$ and $x(i-1) = (\alpha_j, m)$ then

$$\vec{s}^{\upharpoonleft} \stackrel{\text{def}}{=} \langle s_{\alpha_0}, \ldots, s_{\alpha_j} \upharpoonright m, \ldots, s_{\alpha_n} \rangle \in \Sigma,$$

i.e., the "x-predecessor" of $\vec{s}$ is in Σ, and

- there is some $\vec{R} \leq \vec{T}$ such that $\mathsf{tr}_{\vec{R}} = \vec{s}$.

This is the set of the sequences of possible trunks for extensions of $\vec{T}$, whose lengths are ordered via x.

There are three kinds of sequences of possible trunks for extensions of $\vec{T}$; ones that may be involved in forcing $\varphi(\tau_1, \dots, \tau_n)$, ones that may be involved in forcing $\neg\varphi(\tau_1, \dots, \tau_n)$, and the undecided ones. Define such sets formally:

$$\Sigma_0 \stackrel{\text{def}}{=} \{\vec{s} \in \Sigma \,;\, \exists \vec{R} \leq \vec{T}(\vec{s} = \mathsf{tr}_{\vec{R}} \text{ and } \vec{R} \Vdash \varphi(\tau_1, \dots, \tau_n))\},$$

$$\Sigma_1 \stackrel{\text{def}}{=} \{\vec{s} \in \Sigma \,;\, \exists \vec{R} \leq \vec{T}(\vec{s} = \mathsf{tr}_{\vec{R}} \text{ and } \vec{R} \Vdash \neg\varphi(\tau_1, \dots, \tau_n))\},$$

and

$$\Sigma_2 \stackrel{\text{def}}{=} \{\vec{s} \in \Sigma \,;\, \forall \vec{R} \leq \vec{T}(\text{if } \vec{s} = \mathsf{tr}_{\vec{R}} \text{ then } \vec{R} \in \mathbb{P} \text{ does not decide } \varphi(\tau_1, \dots, \tau_n))\}.$$

We have that $\Sigma = \Sigma_0 \cup \Sigma_1 \cup \Sigma_2$. For every $i \in \omega$ define

$$\tilde{S}_i \stackrel{\text{def}}{=} \{\vec{s} \in \Sigma \,;\, \mathsf{dom}^2\vec{s} \setminus \mathsf{dom}^2\vec{t} = x``(i+1)\},$$

the set of all extended trunks in Σ at x-stage i. Note again that $x``(i+1)$ is the part of $c \times \omega$ that needs to be filled so $\vec{t}$ will reach $\vec{s}$.

For every $i \leq j \in \omega$ we define a set of functions $S_{i,j}$ on $\tilde{S}_i$. This is an inductive definition on $j - i$. For $\ell \in 3$ and $\vec{s} \in \tilde{S}_i$ let

- $S_{i,i}(\vec{s}) \stackrel{\text{def}}{=} \ell$ if $\vec{s} \in \Sigma_\ell$, and
- for $i < j$, if $x(i+1) = (\alpha_k, m)$ let $S_{i,j}(\vec{s}) \stackrel{\text{def}}{=} \ell$ if the set

$$\{\beta \,;\, s_{\alpha_k}{}^\frown\langle\beta\rangle \in T_{\alpha_k} \text{ and } S_{i+1,j}((s_{\alpha_0}, \dots, s_{\alpha_k}{}^\frown\langle\beta\rangle, \dots, s_{\alpha_n}) = \ell)\} \in \Phi_{\alpha_k}.$$

Recursively on $i \in \omega$ define $\tilde{S}'_i \subseteq \tilde{S}_i$ as follows. Using the ω-completeness[1] of $\Phi_{\vec{t}}$ we find $\tilde{S}'_0 \subseteq \tilde{S}_0$ homogeneous for all $S_{0,j}$ and such that if $x(0) = (\alpha_k, \beta)$, then $\{\beta \,;\, (t_{\alpha_0}, \dots, t_{\alpha_k}{}^\frown\langle\beta\rangle, \dots, t_{\alpha_n}) \in \tilde{S}'_0\} \in \Phi_{\alpha_k}$. For $0 < i < \omega$ we define

$$\tilde{S}'_i \stackrel{\text{def}}{=} \{\vec{s} \in \tilde{S}_i \,;\, \vec{s}^{\,\upharpoonleft} \in \tilde{S}'_{i-1} \text{ and } \forall i \leq j \in \omega,\ S_{i,j}(\vec{s}) = S_{i-1,j}(\vec{s}^{\,\upharpoonleft})\}.$$

Remember that $\vec{s}^{\,\upharpoonleft}$ is $\vec{s}$ without the x-last element added.

By induction, $\tilde{S}'_i$ is homogeneous for all $S_{i,j}$ for $i \leq j \in \omega$. The definition of the $S_{i,j}$ implies that for each $\vec{r} \in \tilde{S}'_{i-1}$,

$$\{\beta \,;\, (r_{\alpha_0}, \dots, r_{\alpha_k}{}^\frown\langle\beta\rangle, \dots, r_{\alpha_n}) \in \tilde{S}'_i\} \in \Phi_{\alpha_k}. \tag{2}$$

Define $\tilde{S} \stackrel{\text{def}}{=} \{\vec{t}\} \cup \bigcup\{\tilde{S}'_i \,;\, i \in \omega\}$. Before we go on, we check if we are in the right direction.

Claim 1.
If $\vec{R}, \vec{Q} \in E_c$, are such that $\mathsf{tr}_{\vec{R}}, \mathsf{tr}_{\vec{Q}} \in \tilde{S}$, and $\vec{R}, \vec{Q} \leq \vec{T}$ then we ca not have that $\vec{R} \Vdash \varphi(\tau_1, \dots, \tau_n)$ and $\vec{Q} \Vdash \neg\varphi(\tau_1, \dots, \tau_n)$.

[1]This is the only place where we use some sort of completeness of the ultrafilters. This indicates that this sort of Prikry lemma could hold also when forcing with weaker ultrafilters.

PROOF OF CLAIM. Assume the contrary. For some $i, j \in \omega$ we have that $\mathsf{dom}^2(\mathsf{tr}_{\vec{R}}) = \mathsf{dom}^2(\vec{t}) \cup x``i$ and $\mathsf{dom}^2(\mathsf{tr}_{\vec{Q}}) = \mathsf{dom}^2(\vec{t}) \cup x``j$. Without loss of generality assume that $i \leq j$.

If $i < j$ we will increase $\vec{R}$ until its x-distance from $\vec{t}$ is also j. Let $x(i) = (\alpha_k, m)$. The set

$$H \stackrel{\text{def}}{=} \{\beta < \alpha_k \, ; \, (\mathsf{tr}_{R_{\alpha_0}}, \ldots, \mathsf{tr}_{R_{\alpha_k}}{}^\frown\langle\beta\rangle, \ldots, \mathsf{tr}_{R_{\alpha_n}}) \in \tilde{S}'_{i_1}\}$$

is in $\Phi_{\mathsf{tr}_{\vec{R}}}$ by (2). Since $\vec{R} \in \mathbb{P}$, the set

$$H' \stackrel{\text{def}}{=} \{\beta < \alpha_k \, ; \, \mathsf{tr}_{R_{\alpha_k}}{}^\frown\{\beta\} \in R_{\alpha_k}\}$$

of the successors of $\mathsf{tr}_{R_{\alpha_k}}$ in R_{α_k} is in the ultrafilter $\Phi_{\mathsf{tr}_{\vec{R}}}$. Let $\beta \in H \cap H'$,

$$\vec{r} \stackrel{\text{def}}{=} (\mathsf{tr}_{R_{\alpha_0}}, \ldots, \mathsf{tr}_{R_{\alpha_k}}{}^\frown\langle\beta\rangle, \ldots, \mathsf{tr}_{R_{\alpha_n}}), \text{ and}$$
$$R'_{\alpha_k} \stackrel{\text{def}}{=} R_{\alpha_k} \uparrow (\mathsf{tr}_{R_{\alpha_k}}{}^\frown\langle\beta\rangle)$$

Then, $\vec{r} \in \tilde{S}'_{i_1} \subseteq \tilde{S}$ because $\beta \in H$ and the condition

$$\vec{R}' \stackrel{\text{def}}{=} \langle R_{\alpha_0}, \ldots, R'_{\alpha_k}, \ldots, R_{\alpha_n}\rangle$$

is stronger than $\vec{R}$. So $\vec{R}' \Vdash \varphi(\tau_1, \ldots, \tau_n)$, and $\mathsf{dom}^2(\vec{r}) = \mathsf{dom}^2(\vec{t}) \cup x``(i+1)$. This way we keep increasing $i$ until we get $\vec{S} \in E_c$, $\vec{S} \leq \vec{R}$ with $\mathsf{dom}^2(\mathsf{tr}_{\vec{S}}) = \mathsf{dom}^2(t) \cup x``j$.

Now, if $j = 0$ then $\mathsf{tr}_{\vec{S}} = \mathsf{tr}_{\vec{R}} = \mathsf{tr}_{\vec{Q}}$ and by Proporsition 2.14 we get $\vec{R} \parallel \vec{Q}$ which is a contradiction. If $j > 0$ then $\mathsf{tr}_{\vec{S}}, \mathsf{tr}_{\vec{Q}} \in \tilde{S}'_{j-1}$. Since $\vec{S} \leq \vec{T}$ and $\vec{S} \Vdash \varphi(\tau_1, \ldots, \tau_n)$, $\mathsf{tr}_{\vec{S}} \in \Sigma_0$ hence $S_{i-1,i-1}(\mathsf{tr}_{\vec{S}}) = 0$. Similarly, $S_{i-1,i-1}(\vec{Q}) = 1$ contradicting the homogeneity of $\tilde{S}'_{i-1}$ and thus proving the claim. qed claim

Now we decompose $\tilde{S}$ to get appropriate trees with the original trunks $\mathsf{tr}_{\vec{T}}$. For each $\alpha \in c$ let

$$S_\alpha \stackrel{\text{def}}{=} \{s \subseteq \omega \times \alpha \, ; \, \exists \vec{r} \in \tilde{S}(r_\alpha = s)\}.$$

Claim 2.
The sequence $\vec{S} \stackrel{\text{def}}{=} \langle S_\alpha \, ; \, \alpha \in c\rangle$ is a condition in $\mathbb{P}$.

PROOF OF CLAIM. Let $\alpha \in c$. We only must show that for any $s \in S_\alpha$ with $s \trianglerighteq \mathsf{tr}_{T_\alpha}$ the set $\mathsf{Suc}_{S_\alpha}(s)$ is in Φ_α. Let $i \in \omega$ be such that $x(i) = (\alpha, \mathsf{dom}(s))$. Let $\vec{r} \in \tilde{S}'_i$ be such that $r_\alpha = s$. We have that

$$\begin{aligned} \mathsf{Suc}_{S_\alpha}(s) &= \{\beta < \alpha \, ; \, s^\frown\langle\beta\rangle \in S_\alpha\} \\ &= \{\beta < \alpha \, ; \, \exists \vec{r} \in \tilde{S}'_{i+1}, \, r_\alpha = s^\frown\langle\beta\rangle\} \supseteq \\ &\supseteq \{\beta < \alpha \, ; \, (r_{\alpha_0}, \ldots, r_\alpha{}^\frown\langle\beta\rangle, \ldots, r_{\alpha_n}) \in \tilde{S}'_{i+1}\}, \end{aligned}$$

which by (2) is in Φ_α. qed claim

Finally, assume for a contradiction that $\vec{S}$ does not decide $\varphi(\tau_0, \ldots, \tau_n)$. Then there are $\vec{R}, \vec{Q} \leq \vec{T}$ such that $\vec{R} \Vdash \varphi(\tau_0, \ldots, \tau_n)$ and $\vec{Q} \Vdash \neg\varphi(\tau_0, \ldots, \tau_n)$. By the approximation lemma we may assume that $\mathsf{dom}(\vec{R}) = \mathsf{dom}(\vec{Q}) = c$ and we can also choose $\vec{R}, \vec{Q}$ such that $\mathsf{tr}_{\vec{R}}, \mathsf{tr}_{\vec{Q}} \in \tilde{S}'$ (just extend the trunks appropriately). But this contradicts Claim 1. qed

Now we have all the tools to prove some cardinal preservation.

THEOREM 2.19. *For every* $0 < n < \omega$, κ_n *is a cardinal of* $V(G)$.

PROOF. First let $\kappa_n \in \mathsf{Pr}$. Assume for a contradiction that for some $0 < n < \omega$ there is some $\delta < \kappa_n$ and a bijection $f : \delta \to \kappa_n$ in $V(G)$. Let $\dot{f} \in \mathsf{HS}$ be a name for f with support $E_e \in I$. By the approximation lemma there is an E_e-name for this f. We will show that this is impossible by splitting E_e in three parts (a product), one of which has the κ_n-cc, and the other two do not add bounded subsets to κ_n.

Clearly, E_e is isomorphic to $E_{e\cap\kappa_n} \times E_{(e\setminus\kappa_n)\cap\mathsf{Pr}} \times E_{(e\setminus\kappa_n)\cap\mathsf{Co}}$. The partial order $E_{e\cap\kappa_n}$, has the κ_n-cc, and $E_{(e\setminus\kappa_n)\cap\mathsf{Co}}$ does not add new subsets to κ_n since it is a finite support product of Jech collapses, each of which does not add bounded subsets to κ_{n+1}.

So it is left to show that $E_{(e\setminus\kappa_n)\cap\mathsf{Pr}}$ does not add new subsets to κ_n. To do that let us first see that if $\leq^*$ is the relation of direct extension in $E_{(e\setminus\kappa_n)\cap\mathsf{Pr}}$ (i.e., $(\leq) \cap (E_{(e\setminus\kappa_n)\cap\mathsf{Pr}} \times E_{(e\setminus\kappa_n)\cap\mathsf{Pr}})$ between conditions with the same trunks), then $(E_{(e\setminus\kappa_n)\cap\mathsf{Pr}}, \leq^*)$ is κ_n-closed. Let $\gamma < \kappa_n$ and let $\{\vec{S}_\zeta \, ; \, \zeta < \gamma\}$ be a $\leq^*$-descending sequence of elements in $E_{(e\setminus\kappa_n)\cap\mathsf{Pr}}$. Since all ultrafilters involved in the definition of $E_{(e\setminus\kappa_n)\cap\mathsf{Pr}}$ are κ_n-complete, the sequence

$$\langle \bigcap_{\zeta<\gamma} S_{\zeta\xi} \, ; \, \xi \in (e \setminus \kappa_n) \cap \mathsf{Pr} \text{ and } S_{\zeta\xi} \neq \varnothing \rangle$$

is a condition in $E_{(e\setminus\kappa_n)\cap\mathsf{Pr}}$ and it is stronger than all of the $\vec{S}_\zeta$.

With the standard arguments as in Lemma 2.18 for $c = (e \setminus \kappa_n) \cap \mathsf{Pr}$ we can see that $(E_{(e\setminus\kappa_n)\cap\mathsf{Pr}}, \leq)$ does not add bounded subsets to κ_n. So we have now that E_e cannot collapse κ_n. So the function f whose existence we assumed in the beginning of this proof cannot have an E_e-name. Contradiction.

Now assume that $\kappa_n \in \mathsf{Co}$. Similarly we assume for a contradiction that for some $\delta < \kappa_n$, there is a bijection $f : \delta \to \kappa_n$ in $V(G)$. Let $\dot{f} \in \mathsf{HS}$ be a name for f with support E_e. The partial order E_e is isomorphic to

$$E_{e\cap\kappa_n} \times E_{(e\setminus\kappa_n)\cap\mathsf{Co}} \times E_{(e\setminus\kappa_{n+1})\cap\mathsf{Pr}}.$$

As before, $E_{e\cap\kappa_n}$ has the κ-cc, and $E_{(e\setminus\kappa_n)\cap\mathsf{Co}} \times E_{(e\setminus\kappa_{n+1})\cap\mathsf{Pr}}$ does not add new subsets to κ_n. qed

So we managed to construct a model with any desired pattern of regular and singular cardinals of cofinality ω, in the interval $(\omega, \aleph_\omega)$. It is easy to see how to modify this construction to start from any desired regular cardinal $\lambda \leq \kappa_1$ of the ground model. We just have to change the $\mathbb{P}_\alpha$ for $\alpha \in (\omega, \kappa_1)$ to the appropriate Jech collapses for the $\alpha \in (\lambda, \kappa_1)$.

When we built this symmetric model, the strong compactness of the κ_n was destroyed. But being strongly compact implies a lot of combinatorial properties for the κ_n (see [**Kan03**, pages 307-310]) and not everything is lost in $V(G)$. We will take a look on what combinatorial properties have the κ_n left in $V(G)$. Note that [**Jec03**, Theorem 21.2] says that being strongly compact is preserved by small forcing. We can use this to draw strength from our former strongly compacts.

LEMMA 2.20. *In $V(G)$ every cardinal κ in* Pr *is almost Ramsey.*

PROOF. Fix $\kappa \in \mathsf{Pr}$ and let $f : [\kappa]^{<\omega} \to 2$. Let $\dot{f} \in \mathsf{HS}$ be a name for f with support $E_e \in I$. As usual we can see f as a set of ordinals and so apply the approximation lemma to get that this $\dot{f}$ is in fact an E_e-name for f. As in the proof of Theorem 2.19, E_e is isomorphic to $E_{e\cap\kappa} \times E_{(e\setminus\kappa)\cap\mathsf{Pr}} \times E_{(e\setminus\kappa)\cap\mathsf{Co}}$. By [**Jec03**, Theorem 21.2] we get that after forcing with $E_{e\cap\kappa}$, κ is still strongly compact, and thus still measurable. By [**Kan03**, Exercise 7.19] there is a measure one set of Ramsey cardinals for κ, i.e., if G^* is $E_{e\cap\kappa}$-generic and U is a normal measure for κ in $V[G^*]$ then $\{\alpha < \kappa\ ;\ \alpha \text{ is a Ramsey cardinal}\} \in U$. But the existence of an unbounded subset of κ that contains just Ramsey cardinals means that κ is almost Ramsey in $V[G^*]$. Moreover, as in the proof of Theorem 2.19, we can get that $E_{(e\setminus\kappa)\cap\mathsf{Pr}}$ does not add bounded subsets to κ, and neither does $E_{(e\setminus\kappa)\cap\mathsf{Co}}$. So using [**AK08**, Proposition 3] we get that κ is almost Ramsey after forcing with E_e. qed

The amount of choice failing in this symmetric model depends on the value of $f(0)$. If $f(0) = 0$ then $\mathsf{AC}_\omega(\mathcal{P}(\omega))$ fails, and if $f(0) = 1$ then $\mathsf{AC}_{\omega_1}(\mathcal{P}(\omega))$ fails with arguments similar to the proof of Lemma 1.37.

In the following sections we will construct sequences of just successive singular cardinals and there we will see other remaining combinatorial properties of the former strongly compact cardinals. We could always apply the methods of this section to the next ones to allow for regular cardinals in between, in almost any place we want.

4. Longer countable sequences of singular cardinals

In this section we will construct a sequence of successive singular cardinals (of cofinality ω) that has ordertype larger than ω and smaller or equal to $(\omega_1)^V$. In particular we will start from a sequece of strongly compact cardinals, which sequence has ordertype $\rho \in (\omega, \omega_1^V]$, and we will see a way to deal with the ordinals that are above a limit of these strongly compacts and below the next strongly compact. For example, if $\langle \kappa_\xi\ ;\ \xi < \rho\rangle$ is the sequence of strongly compact cardinals, then we will singularise and collapse the cardinals in the interval $(\bigcup_{\alpha<\omega} \kappa_\alpha, \kappa_\omega)$.

Formally, we assume ZFC and we assume that for some ordinal $\rho \in (\omega, \omega_1]$, there is a ρ-long sequence

$$\langle \kappa_\xi\ ;\ 0 < \xi < \rho\rangle,$$

such that for every $0 < \xi < \rho$, κ_ξ is strongly compact. Let η be the limit of this sequence. This model will satisfy the approximation lemma and will contain a sequence of successive singular cardinals of ordertype ρ. Note that in V we have that $\rho < \kappa_1$ and in the symmetric model, κ_1 will become the first uncountable cardinal. Therefore in the symmetric model the sequence of singular cardinals will be countable. For uncountable sequences we will employ Gitik's original methods in a construction presented in the next section.

For the construction of this section we will start with a finite support product of collapses for the infinite ordinals below κ_1 and some sort of injective Prikry-like forcings above κ_1, as in the previous section.

Call Reg^η the set of infinite regular cardinals $\alpha \in (\omega, \eta)$. For an $\alpha \in \mathsf{Reg}^\eta$ we say that α is of

(type 0) If $\alpha \in (\omega, \kappa_1)$.

(type 1) If $\alpha \geq \kappa_1$ and there is a largest $\kappa_\xi \leq \alpha$ (i.e., $\alpha \in [\kappa_\xi, \kappa_{\xi+1})$). Let U_α be a κ_ξ-complete fine ultrafilter over $\mathcal{P}_{\kappa_\xi}(\alpha)$, and let $h_\alpha : \mathcal{P}_{\kappa_\xi}(\alpha) \to \alpha$ be a surjection. If α is inaccessible then let h_α to be moreover a bijection. Define

$$\Phi_\alpha = \{X \subseteq \alpha \; ; \; h_\alpha^{-1}\text{``}X \in U_\alpha\}.$$

This is a uniform κ_ξ-complete ultrafilter over α.

(type 2) If there is no largest strongly compact $\leq \alpha$, then let $\beta_\alpha \stackrel{\text{def}}{=} \bigcup\{\kappa_\zeta \; ; \; \kappa_\zeta < \alpha\}$. Since $\bigcup\zeta$ is a countable limit ordinal, we can get an increasing cofinal function $g : \omega \to \zeta$ in the ground model. Fix such a function and call it g_α. We have that $\beta_\alpha = \bigcup\{\kappa_{g_\alpha(n)} \; ; \; n < \omega\}$, and that the sequence

$$\langle \kappa_{g_\alpha(n)} \; ; \; n < \omega \rangle$$

is ascending. For each $n < \omega$ let $U_{\alpha,n}$ be a fine ultrafilter over $\mathcal{P}_{\kappa_{g_\alpha(n)}}(\alpha)$ and $h_{\alpha,n} : \mathcal{P}_{\kappa_{g_\alpha(n)}}(\alpha) \to \alpha$ be a surjection. Again, if α is inaccessible then each $h_{\alpha,n}$ is taken to be a bijection. Define

$$\Phi_{\alpha,n} \stackrel{\text{def}}{=} \{X \subseteq \alpha \, ; h_{\alpha,n}^{-1}\text{``}X \in U_{\alpha,n}\}.$$

This is a $\kappa_{g_\alpha(n)}$-complete uniform ultrafilter over α.

For a type 0 cardinal α we will add surjections from ω onto α and thus make α countable.

For a type 1 cardinal α we will use injective tree-Prikry forcing to make α of cofinality ω. The inaccessible cardinals in the interval $(\kappa_\xi, \kappa_{\xi+1})$ will be collapsed to the κ_ξ because these forcings will be isomorphic to strongly compact Prikry forcings, precisely as in Proposition 2.6. Since κ_ξ is a limit of inaccessible cardinals, the interval $(\kappa_\xi, \kappa_{\xi+1})$ will collapse to κ_ξ.

To make type 2 cardinals singular is a little more involved. Inspired by Gitik's treatment of the subject, which is presented in the next section, for a type 2 cardinal α we use g_α to pick a countable cofinal sequence of ultrafilters $\langle \Phi_{g_\alpha(n)} \; ; \; n < \omega \rangle$, in order to ensure that β_α, the limit of the strongly compacts below α is not collapsed. To show that the type 2 cardinals are collapsed (to β_α), we also use the fine ultrafilters and do a similar proof as for the type 1 cardinals.

In the end a permutation of each coordinate in itself and the appropriate finite supports will keep just the κ_δ's and their limits from collapsing. So we will end up with an ρ-long sequence of successive singular cardinals.

Let's proceed in the formal definition of the partial order we're going to use. For every $\alpha \in \mathsf{Reg}^\eta$ of type 0 let

$$\mathbb{P}_\alpha \stackrel{\text{def}}{=} \{p : \omega \rightharpoonup \alpha \,;\, |p| < \omega\}.$$

For every $\alpha \in \mathsf{Reg}^\eta$ of type 1 let $\mathbb{P}_\alpha \stackrel{\text{def}}{=} \mathbb{P}^{\mathsf{t}}_{\Phi_\alpha}$ be the injective tree-Prikry forcing with respect to the ultrafilter Φ_α, as defined in Definition 2.1.

For every $\alpha \in \mathsf{Reg}^\eta$ of type 2 let $\mathbb{P}_\alpha$ consist of all $T \subseteq {}^{<\omega}\alpha$ which are trees with respect to $\trianglelefteq$, have trunks, and are such that for every $t \in T$ with $\mathsf{tr}_T \trianglelefteq t$,

$$\mathsf{Suc}_T(t) = \{\beta < \alpha \,;\, t^\frown\langle\beta\rangle \in T\} \in \Phi_{\alpha,n}.$$

This forcing is ordered by inclusion, i.e., $T \leq_{\mathbb{P}_\alpha} S$ iff $T \subseteq S$.

We will force with the partial order

$$\mathbb{P} \stackrel{\text{def}}{=} \prod^{\mathsf{fin}}_{\alpha \in \mathsf{Reg}^\eta} \mathbb{P}_\alpha.$$

We will denote conditions in $\mathbb{P}$ by $\vec{T} \stackrel{\text{def}}{=} \langle T_\alpha \,;\, \alpha \in \mathsf{dom}(\vec{T})\rangle$

For each $\alpha \in \mathsf{Reg}^\eta$, let $\mathcal{G}_\alpha$ be the group of permutations of α that move only finitely many elements of α. Let $\mathcal{G}$ be the finite support product of all these $\mathcal{G}_\alpha$'s. We write $\vec{a} = \langle a_\alpha \,;\, \alpha \in \mathsf{Reg}^\eta\rangle$ for an element of $\mathcal{G}$ and mean that only finitely many of the a_α are not the identity. For $\vec{T} \in \mathbb{P}$ and $\vec{a} = \langle a_\alpha \,;\, \alpha \in \mathsf{Reg}^\eta\rangle \in \mathcal{G}$, define

$$\vec{a}(\vec{T}) \stackrel{\text{def}}{=} \langle a\text{``}\text{``}T_\alpha \,;\, \alpha \in \mathsf{dom}(\vec{T})\rangle.$$

PROPOSITION 2.21. *The map $a : \mathbb{P} \to \mathbb{P}$ is an automorphism.*

PROOF. Let $\vec{T} \in \mathbb{P}$ and $\alpha \in \mathsf{dom}(\vec{T})$. If α is of type 0 then the map $T_\alpha \mapsto a_\alpha\text{``}\text{``}T_\alpha$ is clearly an automorphism of $\mathbb{P}_\alpha$. If α is of type 1 then the map $T_\alpha \mapsto a_\alpha\text{``}\text{``}T_\alpha$ is an automorphism of $\mathbb{P}_\alpha$ as we saw in Proposition 2.9. Similarly to the proof of this proposition we can show that $T_\alpha \mapsto a_\alpha\text{``}\text{``}T_\alpha$ is an automorphism also if α is of type 2. qed

For a finite $e \subseteq \mathsf{Reg}^\eta$, define $E_e \stackrel{\text{def}}{=} \{\vec{T}\restriction e \,;\, \vec{T} \in \mathbb{P}\}$. Take the symmetry generator $I \stackrel{\text{def}}{=} \{E_e \,;\, e \subset \mathsf{Reg}^\eta \text{ is finite}\}$. This symmetry generator is projectable, with projections $\vec{T}\restriction^* E_e = \vec{T}\restriction e$. Let

$$V(G) = V(G)^{\mathcal{G},\mathcal{F}}.$$

Similarly to Proposition 2.14 and Lemma 2.15 we can prove the following two facts.

PROPOSITION 2.22. *Let $\vec{T}, \vec{S} \in \mathbb{P}$ be such that $\mathsf{dom}(\vec{T}) = \mathsf{dom}(\vec{S})$, for each $\alpha \in \mathsf{dom}(\vec{T}) \cap \kappa_1$ we have that $T_\alpha = S_\alpha$, and for each $\alpha \in \mathsf{dom}(\vec{T}) \setminus \kappa_1$ we have that $\mathsf{tr}_{T_\alpha} = \mathsf{tr}_{S_\alpha}$. Then the sequence*

$$\langle T_\alpha \,;\, \alpha \in \mathsf{dom}(\vec{T}) \cap \kappa_1\rangle^\frown\langle T_\alpha \cap S_\alpha \,;\, \alpha \in \mathsf{dom}(\vec{T}) \setminus \kappa_1\rangle$$

is in $\mathbb{P}$ and it is stronger than both $\vec{S}$ and $\vec{T}$. In fact, this holds for $< \kappa_1$-many conditions with the same requirements.

LEMMA 2.23. *For every $e \subset \mathsf{Reg}^\eta$ finite, every $\vec{T} \in \mathbb{P}$ and every $\vec{S} \in \mathbb{P}$ such that $\vec{S} \leq \vec{T}\restriction e$ there is an automorphism $\vec{a} \in \mathsf{fix}E_e$ such that $\vec{a}(\vec{T}) \parallel \vec{S}$.*

Consequently the approximation lemma holds for $V(G)$ and if X is a set of ordinals of $V(G)$, then there is some finite subset e of Reg^η, such that $X \in V[G \cap E_e]$.

To show that all κ_α's and therefore their limits are still cardinals in $V(G)$ we need a version of the Prikry lemma.

LEMMA 2.24. *Let $e \subset \mathsf{Reg}^\eta$ be finite, let $\tau_1, \ldots, \tau_n$ be E_e-names, and let φ be a formula with n-many free variables. Then for every condition $\vec{T} = \langle T_\alpha \,;\, \alpha \in e\rangle \in E_e$ there is a stronger condition $\vec{S} \in E_e$ such that for every $\alpha \in e$ we have that $\mathsf{tr}_{T_\alpha} = \mathsf{tr}_{S_\alpha}$ and $\vec{S}$ decides $\varphi(\tau_1, \ldots, \tau_n)$.*

The proof of this lemma is the same as the proof of Lemma 2.18.

LEMMA 2.25. *For every $0 < \xi < \rho$, κ_ξ is a cardinal in $V(G)$. Consequently, their limits are also preserved.*

PROOF. Assume for a contradiction that for some $0 < \xi < \rho$ there is some $\beta < \kappa_\xi$ and a bijection $f : \beta \to \kappa_\xi$ in $V(G)$. Let $\dot{f} \in \mathsf{HS}$ be a name for f with support E_e for some finite $e \subset \mathsf{Reg}^\eta$. Clearly, $E_e = E_{e\cap\kappa_\xi} \times E_{e\setminus\kappa_\xi}$. The first part, $E_{e\cap\kappa_\xi}$ has the κ_ξ-cc, so it can't add a function like f. Also, similarly to the proof of Lemma 2.19 we can use Lemma 2.24 to get that $E_{e\setminus\kappa_\xi}$ doesn't add bounded subsets to κ_ξ. So such f cannot exist in $V[G \cap E_e]$. qed

LEMMA 2.26. *Every ordinal in $(\mathsf{Reg}^\eta) \setminus \kappa_1$ is singular of cofinality ω in $V(G)$. Thus the interval $[\kappa_1, \eta]$ only contains singular cardinals.*

PROOF. Let α in $\mathsf{Reg}^\eta \setminus \kappa_1$. Then $\bigcup(G \cap E_{\{\alpha\}})$ is the Prikry sequence that is added to α and this has a symmetric name (the canonical name for the generic object restricted to $\{\alpha\}$ supported by $\{\alpha\}$. qed

With the next lemmas we will show that all cardinals in $\mathsf{Reg}^\eta \setminus \{\kappa_\xi \,;\, 0 < \xi < \rho\}$ have collapsed. We will start with the type 1 cardinals.

LEMMA 2.27. *For every $0 < \xi < \rho$ and every $\alpha \in (\kappa_\xi, \kappa_{\xi+1})$, $(|\alpha| = \kappa_\xi)^{V(G)}$.*

PROOF. Let $0 < \xi < \rho$. Since strongly compact cardinals are limits of inaccessible cardinals, it suffices to show that for every inaccessible cardinal $\alpha \in (\kappa_\xi, \kappa_{\xi+1}$, we have that $(|\alpha| = |\kappa_\xi|)^{V(G)}$.

Fix $\alpha \in (\kappa_\xi, \kappa_{\xi+1})$ inaccessible cardinal. Then $h_\alpha : \mathcal{P}_{\kappa_\xi}(\alpha) \to \alpha$ is a bijection. By Proposition 2.5, $\mathbb{P}_\alpha$ is isomorphic to the injective strongly compact tree-Prikry forcing with respect to U_α. Therefore there is a $\mathbb{P}_\alpha$-name for a collapse of α to κ_ξ. Such a name is in HS, supported by $E_{\{\alpha\}}$. qed

Also the regular cardinals of type 2 have collapsed to the singular limits of strongly compacts below them. In this proof we will go more into the details of why this happens instead of using an isomorphism with another partial order.

LEMMA 2.28. *For every α of type 2, if β is the largest limit of strongly compacts below α, then $(|\alpha| = \beta)^{V(G)}$.*

PROOF. Let α be of type 2 and let $\beta = \bigcup_{n\in\omega} \kappa_{g_\alpha(n)}$. Similarly to the proof of the previous lemma we assume that α is inaccessible so each of the $h_{\alpha,n} : \mathcal{P}_{\kappa_{g_\alpha(n)}}(\alpha) \to \alpha$ is a bijection. Let $G_{\{\alpha\}}$ be a generic filter over $E_{\{\alpha\}}$ and look at $V[G_{\{\alpha\}}]$. Let $\langle \alpha_n \,;\, n \in \omega\rangle$ be the Prikry sequence added to α by $E_{\{\alpha\}} \cong \mathbb{P}_\alpha$. We want to show that $\alpha = \bigcup_{n\in\omega} h^{-1}_{\alpha,n}(\alpha_n)$ and that each $h^{-1}_{\alpha,n}(\alpha_n)$ has cardinality $\leq \beta$ in $V[G \cap E_{\{\alpha\}}]$. So then α has collapsed to β in $V[G \cap E_{\{\alpha\}}]$. So we will show that for each $\delta \in \alpha$ there is some $n \in \omega$ such that $\delta \in h^{-1}_{\alpha,n}(\alpha_n)$. Fix $\delta \in \alpha$. For all $n \in \omega$, $U_{\alpha,n}$ is a fine ultrafilter in V so

$$\{A \in \mathcal{P}_{\kappa_{g_\alpha(n)}}(\alpha) \,;\, \delta \in A\} \in U_{\alpha,n}.$$

So by the definition of $\Phi_{\alpha,n}$, for every $n \in \omega$,

$$Z_n \stackrel{\text{def}}{=} \{\zeta \in \alpha \,;\, \delta \in h^{-1}_{\alpha,n}(\zeta)\} \in \Phi_{\alpha,n}.$$

Define the set

$$D_\delta \stackrel{\text{def}}{=} \{T \in E_{\{\alpha\}} \,;\, \exists n \in \mathsf{dom}(\mathsf{tr}_T)(\delta \in h^{-1}_{\alpha,n}(\mathsf{tr}_T(n)))\}.$$

This is dense in $E_{\{\alpha\}}$ and δ was arbitrary. Therefore in $V[G_{\{\alpha\}}]$ we have that $\alpha = \bigcup_{n\in\omega} h^{-1}_{\alpha,n}(\alpha_n)$, a countable union of $\leq \beta$-sized sets. So there is an $E_{\{\alpha\}}$-name for a collapse of α to β, and every $E_{\{\alpha\}}$-name is a name in HS. qed

Since it is clear that all cardinals in (ω, κ_1) are countable in $V(G)$ we have the following.

COROLLARY 2.29. *A cardinal of $V(G)$ in (ω, η) is a successor cardinal in $V(G)$ iff it is in $\{\kappa_\xi \,;\, \xi < \rho\}$. Also, a cardinal of $V(G)$ in (ω, η) is a limit cardinal in $V(G)$ iff it is a limit in the sequence $\langle \kappa_\alpha \,;\, \alpha < \rho\rangle$ in V.*

Now let us look at the combinatorial residue from the strong compactness.

LEMMA 2.30. *In $V(G)$ every cardinal κ in (ω, η) is almost Ramsey, i.e., for every $f : [\kappa]^{<\omega} \to 2$ and every $\alpha < \kappa$ there is a set $H \in [\kappa]^\alpha$ that is homogeneous for f.*

PROOF. If κ is a successor cardinal in $V(G)$, in the interval (ω, η) then by Corollary 2.29 there is a $\xi < \rho$ such that $\kappa = \kappa_\xi$. The proof of Lemma 2.20 shows that κ is almost Ramsey. If κ is a limit cardinal in $V(G)$ in the interval (ω, η), then it is a limit of almost Ramsey cardinals thus it is clearly an almost Ramsey cardinal itself. qed

Finally we give the following observation by Arthur Apter.

LEMMA 2.31. *In $V(G)$, all limit cardinals in Reg^η are Rowbottom and they are carrying Rowbottom filters.*

PROOF. Let $\kappa \in \mathsf{Reg}^\eta$ be a limit cardinal of $V(G)$. Let $\delta < \kappa$, $f : [\kappa]^{<\omega} \to \delta$ be a partition, and $\dot{f} \in \mathsf{HS}$ be a name for f with support $E_e \in I$. Let G_e be a generic filter over E_e. We have that $E_e = E_{e\setminus\kappa} \times E_{e\cap\kappa}$. Since $E_{e\setminus\kappa}$ doesn't add bounded subsets to κ and since $|E_{e\cap\kappa}| < \kappa$, we have that in $V[G_e]$, κ is still a limit of measurable cardinals. Since κ has cofinality ω, let $\langle \kappa'_n \,;\, n \in \omega\rangle$ be a sequence of measurable cardinals that is cofinal in κ. For every $n \in \omega$ let U_n be a normal measure over κ'_n.

In $V[G_e]$ define the following sets. For each $n \in \omega$,

$$\bar{U}_n \stackrel{\text{def}}{=} \{X \subseteq \kappa'_n ; \exists Y \subseteq U_n (Y \subseteq X)\}, \text{ and}$$
$$F \stackrel{\text{def}}{=} \{A \subseteq \kappa ; \exists m \forall n \geq m (A \cap \kappa_n \in \bar{U}_n)\}.$$

By the proof of [**Kan03**, Theorem 8.7], F is a Rowbottom filter for κ, so there is a homogeneous set for f in $V[G_e] \subseteq V(G)$. But this shows that if we define F in $V(G)$ then it is a Rowbottom filter for κ in $V(G)$. qed

As a corollary we get the following.

COROLLARY 2.32. *If we assume that we started from a model where all strongly compacts are limits of measurable cardinals, then in $V(G)$ all cardinals in the interval (ω, η) are Rowbottom and they are carrying Rowbottom filters.*

5. Uncountably long sequences of successive singulars with a measurable on top

If we want to construct longer sequences of singular cardinals of cofinality ω, in particular sequences of length strictly greater than ω_1 then we fall back to a construction closer to Gitik's original construction, in order to deal with the (uncountable) limits of strongly compact cardinals. There is a way to do Gitik's construction with a forcing more similar to the one of the previous sections, i.e., by adding trees. In such a forcing, for the type 2 cardinals one would have to add certain names for trees, for the cardinals after an uncountable sequence of strongly compact cardinals. We found that such an iteration becomes more complex than the original construction by Gitik, therefore our modification is based on Gitik's original presentation.

This section is also submitted for publication as a joint paper with Arthur Apter and Peter Koepke, entitled "The first measurable cardinal can be the first uncountable regular cardinal at any successor height" [**ADK**]. In fact we can also have this measurable at certain limit stages. We will turn a sequence $\langle \kappa_i \; ; \; 1 \leq i < \rho \rangle$ of strongly compact cardinals that has no regular limits, into the sequence $\langle \omega_{i+1} \; ; \; 1 \leq i < \rho \rangle$. Moreover, we assume that in the ground model there is a measurable cardinal $\kappa_\rho > \bigcup_{1 \leq i < \rho} \kappa_i$, who is turned into the successor of $(\bigcup_{1 \leq i < \rho} \kappa_i)^V$. This will yield a model of ZF in which the first measurable and first regular cardinal is simultaneously $\aleph_{\rho+1}$. If we assume that $\kappa_\rho = \bigcup_{1 \leq i \leq \rho} \kappa_i$ is a measurable cardinal and we skip the last part of the forcing below, then we will end up with a model where κ_ρ is the first measurable, it is a regular limit cardinal, and every infinite cardinal below it has cofinality ω.

As we mentioned in the introduction, there has been research on the topic of a measurable cardinal being both the least measurable and the least regular uncountable cardinal. Apter in [**Apt96**] (see also comments in the introduction) showed that the consistency strength of $\aleph_2$, $\aleph_{\omega+1}$, or $\aleph_{\omega+2}$ being such a cardinal, has AD as an upper bound. In this section we are able to make any successor cardinal we like the first measurable and first regular uncountable cardinal.

5.1. The Gitik construction. For this construction only, we will go at some point to Boolean-valued models and symmetric submodels of them. This method is presented in [**Jec03**, Chapter 15].

Let $\rho \in \mathsf{Ord}$. We start with an increasing sequence of cardinals,

$$\langle \kappa_\xi \ ; \ \xi < \rho \rangle,$$

such that for every $\xi < \rho$, κ_ξ is strongly compact, and such that the sequence has no regular limits. Let $\kappa_\rho > \bigcup_{\xi<\rho} \kappa_\xi$ be a measurable cardinal.

Call $\mathsf{Reg}^{\kappa_\rho}$ the set of regular cardinals $\alpha \in [\omega, \kappa_\rho)$ in V. For convenience call $\omega \stackrel{\text{def}}{=} \kappa_{-1}$. For an $\alpha \in \mathsf{Reg}^{\kappa_\rho}$ we define a $\mathsf{cf}'\alpha$ to distinguish between the following categories.

(type 1) If there is a largest $\kappa_\xi \leq \alpha$ (i.e., $\alpha \in [\kappa_\xi, \kappa_{\xi+1})$). We then define $\mathsf{cf}'\alpha \stackrel{\text{def}}{=} \alpha$.

If $\alpha = \kappa_\xi$ and $\xi \neq -1$, then let Φ_{κ_ξ} be a measure for κ_ξ. If $\alpha = \omega$ then let Φ_ω be any uniform ultrafilter on ω.

If $\alpha > \kappa_\xi$ is inaccessible, then let H_α be a fine ultrafilter over $\mathcal{P}_{\kappa_\xi}(\alpha)$ and let $h_\alpha : \mathcal{P}_{\kappa_\xi}(\alpha) \to \alpha$ be a bijection. Define

$$\Phi_\alpha := \{X \subseteq \alpha \ ; \ h_\alpha^{-1}``X \in H_\alpha\}.$$

This is a uniform κ_ξ-complete ultrafilter over α.

If $\alpha > \kappa_\xi$ is not inaccessible, then let Φ_α be any κ_ξ-complete uniform ultrafilter over α.

(type 2) If there is no largest strongly compact $\leq \alpha$. We then let β be the largest (singular) limit of strongly compacts $\leq \alpha$. Define $\mathsf{cf}'\alpha \stackrel{\text{def}}{=} \mathsf{cf}\beta$. Let

$$\langle \kappa_\nu^\alpha \ ; \ \nu < \mathsf{cf}'\alpha \rangle$$

be a fixed ascending sequence of strongly compacts $\geq \mathsf{cf}'\alpha$ such that $\beta = \bigcup\{\kappa_\nu^\alpha \ ; \ \nu < \mathsf{cf}'\alpha\}$.

If α is inaccessible, then for each $\nu < \mathsf{cf}'\alpha$, let $H_{\alpha,\nu}$ be a fine ultrafilter over $\mathcal{P}_{\kappa_\nu^\alpha}(\alpha)$ and $h_{\alpha,\nu} : \mathcal{P}_{\kappa_\nu}(\alpha) \to \alpha$ a bijection. Define

$$\Phi_{\alpha,\nu} \stackrel{\text{def}}{=} \{X \subseteq \alpha \, ; \, h_\alpha^{-1}``X \in H_{\alpha,\nu}\}.$$

Again, $\Phi_{\alpha,\nu}$ is a κ_ν^α-complete uniform ultrafilter over α.

If α is not inaccessible, then for each $\nu < \mathsf{cf}'\alpha$, let $\Phi_{\alpha,\nu}$ be any κ_ν^α-complete uniform ultrafilter over α.

This $\mathsf{cf}'\alpha$ will be used when we want to organise the choice of ultrafilters for the type 2 cardinals, since here we cannot use functions g_α like in the definitions of the previous section to do that.

Note that unlike the previous section, there are no type 0 cardinals here. Instead, the infinite regular cardinals below κ_0 are now included in the type 1 cardinals. This is because of the interweaved nature of the forcing here. In particular, to prove that certain cardinals do not collapse we needed to have a sort of homogeneous forcing. This is also the reason why we cannot simply extend the methods of Section ?? here.

The forcing at type 1 cardinals is similar to the one in the previous sections. To singularise type 2 ordinals Gitik used a technique he credits in [**Git80**] to Magidor, a Prikry-type forcing that relies on the countable cofinal sequence $\vec{c}$ that we build for $\mathsf{cf}'\alpha$ to pick a countable sequence of ultrafilters $\langle \Phi_{\vec{c}(n)}\ ;\ n \in \omega\rangle$. This is similar to the forcing we used in the previous section for type 2 cardinals but it is based on $\mathsf{cf}'\alpha$ instead of the functions g_α. To show that the type 2 cardinals are collapsed, we use again the fine ultrafilters, just as we did in the previous section.

As usual we will prove a Prikry-like lemma (see also [**Git80**, Lemma 5.1]). For the arguments in that proof one requires the forcing conditions to grow nicely.

These conditions can be viewed as trees. These trees will grow from "left to right" in order to ensure that a type 2 cardinal α will have the necessary information from the Prikry sequence at stage $\mathsf{cf}'\alpha$. Let's take a look at the definition of the stems of the Prikry sequences to be added.

DEFINITION 2.33. For $t \subseteq \mathsf{Reg}^{\kappa_\rho} \times \omega \times \kappa_\rho$ we define the sets

$$\mathsf{dom}(t) \stackrel{\text{def}}{=} \{\alpha \in \mathsf{Reg}^{\kappa_\rho}\ ;\ \exists m \in \omega \exists \gamma \in \mathsf{Ord}((\alpha, m, \gamma) \in t)\}, \text{ and}$$
$$\mathsf{dom}^2(t) \stackrel{\text{def}}{=} \{(\alpha, m) \in \mathsf{Reg}^{\kappa_\rho} \times \omega\ ;\ \exists \gamma \in \mathsf{Ord}((\alpha, m, \gamma) \in t)\}.$$

Let P_1 be the set of all finite subsets t of $\mathsf{Reg}^{\kappa_\rho} \times \omega \times \kappa_\rho$, such that for every $\alpha \in \mathsf{dom}(t)$, $t(\alpha) \stackrel{\text{def}}{=} \{(m, \gamma)\ ;\ (\alpha, m, \gamma) \in t\}$ is an injective function from some finite subset of ω into α.

To add a Prikry sequence to a type 2 cardinal α, we want to have some information on the Prikry sequence of the cardinal $\mathsf{cf}'\alpha$. We also want to make sure that these stems are appropriately ordered for the induction in the proof of the aforementioned Prikry-like lemma. So we define the following.

DEFINITION 2.34. Let P_2 be the set of all $t \in P_1$ such that

(1) for every $\alpha \in \mathsf{dom}(t)$, $\mathsf{cf}'\alpha \in \mathsf{dom}(t)$ and $\mathsf{dom}(t(\mathsf{cf}'\alpha)) \supseteq \mathsf{dom}(t(\alpha))$, and
(2) if $\{\alpha_0, \ldots, \alpha_{n-1}\}$ is an increasing enumeration of $\mathsf{dom}(t) \setminus \kappa_0$, then there are $m, j \in \omega$, such that $m \geq 1$, $j \leq n-1$ with the properties that
 - for every $k < j$ we have that $\mathsf{dom}(t(\alpha_k)) = m + 1$, and
 - for every $k \in \{j, \ldots, n-1\}$ we have that $\mathsf{dom}(t(\alpha_k)) = m$.

These m and α_j are unique for t and are denoted by $m(t) \stackrel{\text{def}}{=} m$ and $\alpha(t) \stackrel{\text{def}}{=} \alpha_j$. We may think of the point $(\alpha(t), m(t))$ as the point we have to fill in next, in order to extend t.

Let's call elements of P_2 stems. In the following image we can see roughly what a stem t with a domain $\{\alpha_0, \alpha_1, \alpha_2\}$ above κ_0 looks like.

In order to add Prikry sequences, we will use the ultrafilters and define the partial ordering with which we will force.

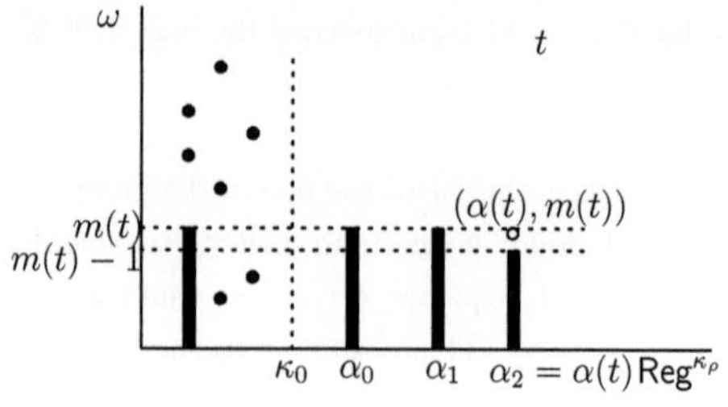

DEFINITION 2.35. Let P_3 be the set of pairs (t, T) such that

(1) $t \in P_2$,
(2) $T \subseteq P_2$,
(3) $t \in T$,
(4) for every $t' \in T$ we have $t' \supseteq t$ or $t' \subseteq t$, and $\mathsf{dom}(t') = \mathsf{dom}(t)$,
(5) for every $t' \in T$, if $t' = r \cup \{(\alpha(r), m(r), \beta)\}$ then $t'^{-} \stackrel{\text{def}}{=} r \in T$, i.e., T is tree-like,
(6) for every $t' \in T$ with $t' \supseteq t$, if $\alpha(t')$ is of type 1 (i.e., $\mathsf{cf}'(\alpha(t')) = \alpha(t')$) then

$$\mathsf{Suc}_T(t') \stackrel{\text{def}}{=} \{\beta\,;\, t' \cup \{(\alpha(t'), m(t'), \beta)\} \in T\} \in \Phi_{\alpha(t')}, \text{ and}$$

(7) for every $t' \in T$ with $t' \supseteq t$, if $\alpha(t')$ is of type 2 (i.e., $\mathsf{cf}'\alpha(t') < \alpha(t')$) and $m(t') \in \mathsf{dom}(t'(\mathsf{cf}'(\alpha(t'))))$ then

$$\mathsf{Suc}_T(t') \stackrel{\text{def}}{=} \{\beta\,;\, t' \cup \{(\alpha(t'), m(t'), \beta)\} \in T\} \in \Phi_{\alpha(t'), t'(\mathsf{cf}'\alpha(t'))(m(t'))}.$$

For a (t, T) in P_3 and a subset $x \subseteq \mathsf{Reg}^{\kappa_\rho}$ we write $T \restriction x$ for $\{t' \restriction x\,;\, t' \in T\}$.

We call t the trunk of (t, T).

This P_3 is the forcing we are going to use. It is partially ordered by $(t, T) \leq (s, S) :\Longleftrightarrow$

$$t \restriction \kappa_0 \supseteq s \restriction \kappa_0,\ T \restriction (\mathsf{dom}(s) \setminus \kappa_0) \subseteq S, \text{ and } \mathsf{dom}(t) \supseteq \mathsf{dom}(s).$$

Let $\mathcal{G}$ be the group of permutations of $\mathsf{Reg}^{\kappa_\rho} \times \omega \times \kappa_\rho$ whose elements a satisfy the following properties.

- For every $\alpha \in \mathsf{Reg}^{\kappa_\rho}$ there is a permutation a_α of α that moves only finitely many elements of α, and is such that for each $n \in \omega$ and each $\beta \in \alpha$,

$$a(\alpha, n, \beta) = (\alpha, n, a_\alpha(\beta)).$$

  The finite subset of α that a_α moves, we denote by $\mathsf{supp}(a_\alpha)$, which stands for "support of a_α".
- For only finitely many $\alpha \in \mathsf{Reg}^{\kappa_\rho}$ is a_α not the identity. This finite subset of $\mathsf{Reg}^{\kappa_\rho}$ we denote by $\mathsf{dom}(a)$.

We extend $\mathcal{G}$ to P_3 as follows. For $a \in \mathcal{G}$ and $(t, T) \in P_3$, define

$$a(t, T) \stackrel{\text{def}}{=} (a``t, \{a``t'\,;\, t' \in T\}),$$

where $a``t \stackrel{\text{def}}{=} \{(\alpha, n, a_\alpha(\beta))\ ;\ (\alpha, n, \beta) \in t\}$.

Unfortunately, in general $a(t, T)$ is not a member of P_3 because of the branching condition at type 2 cardinals. In particular, it is possible that for some $\alpha \in \mathsf{dom}(t)$ of type 2, and some $t' \in T$ with $\alpha = \alpha(t')$, we have that $a_{\mathsf{cf}'\alpha}(t'(\mathsf{cf}'\alpha)(m(t'))) = \gamma \neq t'(\mathsf{cf}'\alpha)(m(t'))$, and even though we had before $\mathsf{Suc}_T(t') \in \Phi_{\alpha, t'(\mathsf{cf}'\alpha)(m(t'))}$, it is not true that $\mathsf{Suc}_T(t') \in \Phi_{\alpha,\gamma}$.

To overcome this problem, for an $a \in \mathcal{G}$, define $P^a \subseteq P_3$ as follows.

(t, T) is in P^a iff the following hold:

(1) $\mathsf{dom}(t) \supseteq \mathsf{dom}(a)$,
(2) for every $\alpha \in \mathsf{dom}(t)$ we have that $\mathsf{dom}(t(\alpha)) = \mathsf{dom}(t(\mathsf{cf}'\alpha))$, and
(3) for every $\alpha \in \mathsf{dom}(t)$, we have that

$$\mathsf{rng}(t(\alpha)) \supseteq \{\beta \in \mathsf{supp}(a_\alpha)\ ;\ \exists q \in T(\beta \in \mathsf{rng}(q(\alpha)))\}.$$

The equality in (2) ensures the requirements for membership in ultrafilters of the form $\Phi_{\alpha,\gamma}$. In (3) we require that the stem of each condition already contains all the ordinals that the a_α could move. This will prevent again any trouble with membership in the ultrafilters. One may think that this requirement should be $\mathsf{supp}(a_\alpha) \subseteq \mathsf{rng}(t(\alpha))$ but this is not the case; note that there might be some γ in a_α which doesn't appear in the range of any $q \in T$.

Now, we have that $a : P^a \to P^a$ is an automorphism. Also, as mentioned in [**Git80**, page 68], for every $a \in \mathcal{G}$ the set P^a is a dense subset of P_3. Therefore, a can be extended to a unique automorphism of the complete Boolean algebra B. This is the reason why we go to Boolean valued models here; our symmetric forcing technique cannot deal with such cases. It's a future project to incorporate this case in symmetric forcing with partial orders.

We denote the automorphism of B with the same letter a, and also by $\mathcal{G}$ the automorphism group of B that consists of these extended automorphisms. By [**Jec03**, (14.36)], every automorphism a of B induces an automorphism of the Boolean valued model V^B.

Proceeding to the definition of our symmetric model, for every $e \subseteq \mathsf{Reg}^{\kappa_\rho}$ define

$$E_e \stackrel{\text{def}}{=} \{(t, T) \in P_3\ ;\ \mathsf{dom}(t) \subseteq e\},$$

$$I \stackrel{\text{def}}{=} \{E_e\ ;\ e \subset \mathsf{Reg}^{\kappa_\rho} \text{ is finite and closed under } \mathsf{cf}'\},$$

$$\mathsf{fix}E_e \stackrel{\text{def}}{=} \{a \in \mathcal{G}\ ;\ \forall \alpha \in e(a_\alpha \text{ is the identity on } \alpha)\},$$

and let $\mathcal{F}$ be the normal filter over $\mathcal{G}$ that is generated by

$$\{\mathsf{fix}E_e\ ;\ E_e \in I\}.$$

For each $\dot{x}$ in the Boolean valued model V^B, we define its symmetry group as usual by

$$\mathsf{sym}(\dot{x}) \stackrel{\text{def}}{=} \{a \in \mathcal{G}\ ;\ a(\dot{x}) = \dot{x}\},$$

and we call a name $\dot{x}$ symmetric iff its symmetry group is in the filter $\mathcal{F}$. The class of hereditarily symmetric names HS is defined just as for symmetric forcing with partial orders. We will say that an $E_e \in I$ supports a name $\dot{x} \in \mathsf{HS}$ if $\mathsf{fix}E_e \subseteq \mathsf{sym}(\dot{x})$.

For some V-generic ultrafilter G on B we define the symmetric model

$$V(G) \overset{\text{def}}{=} \{\dot{x}^G \,;\, \dot{x} \in \mathsf{HS}\}.$$

By [**Jec03**, Lemma 15.51], this is a model of ZF, and $V \subseteq V(G) \subseteq V[G]$.

For each $(t,T) \in P_3$ and each $E_e \in I$, define

$$(t,T) \restriction^* E_e \overset{\text{def}}{=} (t \restriction e, \{t' \restriction e \,;\, t' \in T\}).$$

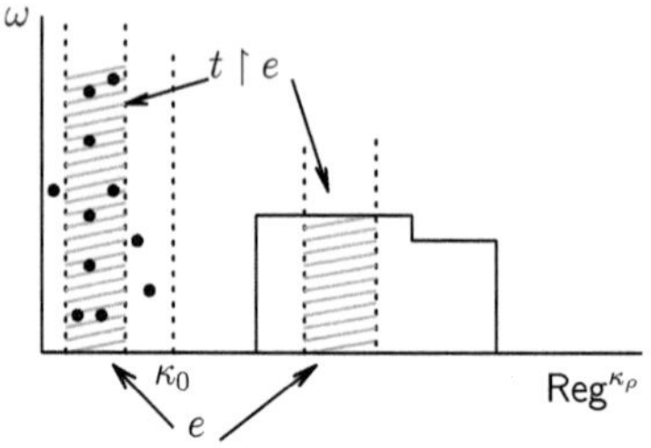

According to [**Git80**, Lemma 3.3.], if φ is a formula with n free variables, $\dot{x}_1, \ldots, \dot{x}_n \in \mathsf{HS}$, and $E_e \in I$ is such that $\mathsf{sym}(\dot{x}_1), \ldots, \mathsf{sym}(\dot{x}_n) \supseteq \mathsf{fix} E_e$ then we have that for every $(t,T) \in P_3$

$$(t,T) \Vdash \varphi(\dot{x}_1, \ldots, \dot{x}_n) \iff (t,T) \restriction^* E_e \Vdash \varphi(\dot{x}_1, \ldots, \dot{x}_n).$$

This implies the approximation lemma.

LEMMA 2.36. *If $X \in V(G)$ is a set of ordinals, then there is an $E_e \in I$ such that $X \in V[G \cap E_e]$.*

PROOF. Because of [**Git80**, Lemma 3.3.] mentioned above, because of the symmetry lemma [**Jec03**, (15.41)], and because canonical names $\check{\alpha}$ are not moved by automorphisms of B, we have that if $\dot{X} \in \mathsf{HS}$ is a P_3-name for X and $E_e \in I$ supports $\dot{X}$ then the set

$$\ddot{X} \overset{\text{def}}{=} \{(\check{\alpha}, (t,T) \restriction^* E_e) \,;\, (t,T) \Vdash \check{\alpha} \in \dot{X}\}$$

is an E_e-name for X. qed

We will use the approximation lemma in all our subsequent proofs.

THEOREM 2.37. *For every $0 \leq \xi \leq \rho$, κ_ξ is a cardinal in $V(G)$. Consequently, their (singular) limits are also preserved.*

PROOF. Assume for a contradiction that there is some $\delta < \kappa_\xi$ and a bijection $f : \delta \to \kappa_\xi$ in $V(G)$. Let $\dot{f}$ be a name for f with support $E_e \in I$. Note that e is a finite subset of $\mathsf{Reg}^{\kappa_\rho}$ that is closed under cf'. By the approximation lemma there is an E_e-name for f, i.e., for this e,

$$f \in V[G \restriction^* E_e].$$

We will show that this is impossible, by taking a dense subset of E_e and showing that it is forcing equivalent to an iterated forcing construction the first part of which has the κ_ξ-c.c., and the other does not collapse κ_ξ (by not adding bounded subsets to κ_ξ, similarly to Prikry forcing).

Claim 1. The following set is dense in E_e.

$$J \stackrel{\text{def}}{=} \{(t,T) \in E_e \,;\, \forall q \in T \forall \alpha \geq \kappa_\xi \forall n < \omega(\text{ if } (\alpha,n) \in \mathsf{dom}^2(q) \setminus \mathsf{dom}^2(t) \text{ and } \mathsf{cf}'\alpha < \alpha \text{ then the ultrafilter } \Phi_{\alpha,q(\mathsf{cf}'\alpha)(n)} \text{ is } \kappa_\xi\text{-complete})\}$$

Proof of claim. First notice that the set

$$D \stackrel{\text{def}}{=} \{(t,T) \in E_e \,;\, \forall \alpha \in \mathsf{dom}(t)(\mathsf{dom}(t(\alpha)) = \mathsf{dom}(t(\mathsf{cf}'\alpha)))\}$$

is dense in E_e. This proof is similar to the proof that for $a \in \mathcal{G}$, P^α is dense in P_3. Now we will prove that for every $(t,T) \in D$ there is a $T' \subseteq T$ such that $(t,T') \in J$. For every $\alpha \in \mathsf{dom}(t) \setminus \kappa_\xi$ such that $\mathsf{cf}'\alpha < \alpha$, let λ_α be the least ordinal $\nu < \mathsf{cf}'\alpha$ such that $\kappa_\nu^\alpha \geq \kappa_\xi$.

We have that $(t,T') \in J$ iff for all $q \in T'$, if $\alpha(q) \geq \kappa_\xi$ and $\mathsf{cf}'\alpha(q) < \alpha(q)$ then $q(\mathsf{cf}'\alpha(q))(m(q)) \geq \lambda_{\alpha(q)}$. This equivalence is true because the right hand side of the implication above ensures that the ultrafilter $\Phi_{\alpha,q(\mathsf{cf}'\alpha),m(q)}$ is κ_ξ-complete. Define

$$b \stackrel{\text{def}}{=} \{\mathsf{cf}'\alpha \,;\, \alpha \in \mathsf{dom}(t) \setminus \kappa_\xi\},$$

and for $\beta \in b$ define

$$c_\beta \stackrel{\text{def}}{=} \max\{\lambda_\alpha \,;\, \alpha \in \mathsf{dom}(t) \setminus \kappa_\xi \text{ and } \mathsf{cf}'\alpha = \beta\}.$$

Then $(t,T') \in J$ if for all $q \in T'$ and $(\alpha,m) \in \mathsf{dom}^2(q) \setminus \mathsf{dom}^2(t)$ we have that $q(\alpha')(m) \geq c_{\alpha'}$. So let

$$T' \stackrel{\text{def}}{=} \{q \in T \,;\, \forall \alpha' \in b \forall m < \omega(\text{if } (\alpha',m) \in \mathsf{dom}^2(q) \setminus \mathsf{dom}^2(t) \text{ then } q(\alpha')(m) \geq c_{\alpha'})\}.$$

Clearly, this $(t,T') \in J$. qed claim 1

Without loss of generality assume that $e \cap \kappa_\xi \neq \varnothing$. Define the sets

$$\mathbb{E} \stackrel{\text{def}}{=} \{(t,T) \upharpoonright^* E_{e\cap\kappa_\xi} \,;\, (t,T) \in J\}, \text{ and}$$
$$P_2^* \stackrel{\text{def}}{=} \{t \upharpoonright (e \setminus \kappa_\xi) \,;\, t \in P_2\}.$$

For $s \in P_2^*$ we can define $\alpha(s)$ and $m(s)$ as we did for the $s \in P_2$, in (2) of the definition of P_2.

Let G^* be an $\mathbb{E}$-generic filter and note that for every $\alpha \in e \setminus \kappa_\xi$ such that $\mathsf{cf}'\alpha < \kappa_\xi$, the set $\langle \bigcup G * (\mathsf{cf}'\alpha)(m) \,;\, m \in \omega\rangle$ is the Prikry sequence that is added to $\mathsf{cf}'\alpha$ by $\mathbb{E}$.

In $V[G^*]$ we define a partial order $\mathbb{Q}$ by $(s,S) \in \mathbb{Q} :\Longleftrightarrow$

(1) $s \in P_2^*$,
(2) $S \subseteq P_2^*$,
(3) $s \in S$,

(4) for all $s' \in S$, $\mathsf{dom}(s') = \mathsf{dom}(s) = e \setminus \kappa_\xi$, and either $s' \supseteq s$ or $s' \subseteq s$,

(5) for every $s' \in S$ with $s' \supseteq s$, if $\alpha(s')$ is of type 1 then

$$\{\beta\ ;\ s' \cup \{(\alpha(s'), m(s'), \beta)\} \in S\} \in \Phi_{\alpha(s')},$$

(6) for every $s' \in S$ with $s' \supseteq s$, if $\alpha(s')$ is of type 2 and $\mathsf{cf}'\alpha(s') \geq \kappa_\xi$ then

$$\{\beta\ ;\ s' \cup \{(\alpha(s'), m(s'), \beta)\} \in S\} \in \Phi_{\alpha(s'), s'(\mathsf{cf}'\alpha(s'))(m(s'))},$$

(7) for every $s' \in S$ with $s' \supseteq s$, if $\alpha(s')$ is of type 2 and $\mathsf{cf}'\alpha(s') < \kappa_\xi$ then

$$\{\beta\ ;\ s' \cup \{(\alpha(s'), m(s'), \beta)\} \in S\} \in \Phi_{\alpha(s'), \bigcup G^*(\mathsf{cf}'\alpha(s'))(m(s'))}, \text{ and}$$

(8) for every $s' \in S$ and every $s'' \in P_2^*$ if $s'' \subseteq s$ then $s'' \in S$, i.e., S is tree-like.

This definition means that $\mathbb{Q}$ is like P_3 but restricted above κ_ξ.

An obvious name for $\dot{\mathbb{Q}}$ for $\mathbb{Q}$ is the following. For $(t, T) \in \mathbb{E}$, $(\sigma, (t, T)) \in \dot{\mathbb{Q}}$ iff

(a) there is an $s \in P_2^*$ and an $\mathbb{E}$-name $\bar{\sigma}$ such that $s \cup t \in P_2$, $\sigma = \mathsf{op}(\check{s}, \bar{\sigma})$, and for all $\pi \in \mathsf{dom}(\bar{\sigma})$ there is a $s' \in P_2^*$ such that $\check{s}' = \pi$.

(b) $(t, T) \Vdash \check{s} \in \bar{\sigma}$,

(c) $(t, T) \Vdash \forall \pi (\pi \in \bar{\sigma} \rightarrow \mathsf{dom}(\pi) = \mathsf{dom}(\check{s}) \wedge (\pi \subseteq \check{s} \vee \pi \supseteq \check{s}))$,

(d) $(t, T) \Vdash \forall \pi (\pi \in \bar{\sigma} \wedge \alpha(\pi) = \mathsf{cf}'\alpha(\pi) \rightarrow \exists \check{X} (\check{X} \in \check{\Phi}_{\alpha(\pi)} \wedge \forall \check{\beta} (\beta \in \check{X} \leftrightarrow \pi \cup \{(\check{\alpha}(\pi), \check{m}(\pi), \check{\beta})\} \in \bar{\sigma})))$,

(e) $(t, T) \Vdash \forall \pi (\pi \in \bar{\sigma} \wedge \alpha(\pi) > \mathsf{cf}'\alpha(\pi) \wedge \mathsf{cf}'\alpha(\pi) \geq \kappa_\xi \rightarrow$
$\exists \check{X} (\check{X} \in \check{\Phi}_{\alpha(\pi), \pi(\mathsf{cf}'\alpha(\pi))(m(\pi))} \wedge \forall \check{\beta} (\beta \in \check{X} \leftrightarrow \pi \cup \{(\check{\alpha}(\pi), \check{m}(\pi), \check{\beta})\} \in \bar{\sigma})))$,

(f) $(t, T) \Vdash \forall \pi (\pi \in \bar{\sigma} \wedge \alpha(\pi) > \mathsf{cf}'\alpha(\pi) \wedge \mathsf{cf}'\alpha(\pi) < \kappa_\xi \rightarrow$
$\exists \check{X} (\check{X} \in \check{\Phi}_{\alpha(\pi)\Gamma(\mathsf{cf}'\alpha(\pi))(m(\pi))} \wedge \forall \check{\beta} (\beta \in \check{X} \leftrightarrow \pi \cup \{(\check{\alpha}(\pi), \check{m}(\pi), \check{\beta})\} \in \bar{\sigma})))$, where Γ is the standard $\mathbb{E}$-name for $\bigcup G^*$,

(g) $(t, T) \Vdash \forall \pi \forall \pi' (\pi \in \bar{\sigma} \wedge \pi' \in \check{P}_2^* \wedge \pi' \subseteq \pi \rightarrow \pi' \in \bar{\sigma})$.

The name for the ordering on $\dot{\mathbb{Q}}$ is defined as

$$(\mathsf{op}(\sigma, \tau), (t, T)) \in \leq_{\dot{\mathbb{Q}}} :\Longleftrightarrow (t, T) \Vdash \sigma \subseteq \tau.$$

From the forcing theorem we have that $\dot{\mathbb{Q}}^{G^*} = \mathbb{Q}$. For every $(t, T) \in P_3$ and $t' \in T$ such that $t' \supseteq t$ define

$$(t, T) \uparrow (t') \stackrel{\text{def}}{=} (t', \{t'' \in T\ ;\ t'' \subseteq t' \text{ or } t \subseteq t''\}),$$

the extension of (t, T) with trunk t'. If $t' \subseteq t$ then we conventionally take $(t, T) \uparrow (t') \stackrel{\text{def}}{=} (t, T)$.

Define a map $i : J \rightarrow \mathbb{E} * \dot{\mathbb{Q}}$. For $(r, R) \in J$, we take $i((r, R)) = ((r_1, R_1), \rho) :\Longleftrightarrow$

(i) $(r_1, R_1) \stackrel{\text{def}}{=} (r, R) \upharpoonright^* E_{e \cap \kappa_\xi}$,

(ii) $\rho = \mathsf{op}(\check{r}_2, \bar{\rho})$, where $r_2 \stackrel{\text{def}}{=} r \upharpoonright (e \setminus \kappa_\xi)$ and for all $\pi \in \mathsf{dom}(\bar{\rho})$, there is an $r' \in R$ such that $(\pi = (r' \upharpoonright (e \setminus \kappa_\xi))^\vee$,

(iii) For all $r' \in R$ with $r' \subsetneq r$ we have that $((r' \upharpoonright (e \setminus \kappa_\xi))^\vee, (r_1, R_1)) \in \bar{\rho}$,

(iv) For all $r' \in R$ with $r' \supseteq r$ we have that

$$((r' \restriction (e \setminus \kappa_\xi))^\vee, (r_1, R_1) \uparrow (r' \restriction (e \cap \kappa_\xi))) \in \bar{\rho}.$$

(v) No other elements are in $\bar{\rho}$.

Claim 2. For all $(r, R) \in J$, $i((r, R)) = ((r_1, R_1), \rho) \in \mathbb{E} * \dot{\mathbb{Q}}$.

Proof of claim. That $(r_1, R_1) \in \mathbb{E}$ is immediate. So we must show that $(r_1, R_1) \Vdash \rho \in \dot{\mathbb{Q}}$. Requirement (a) clearly holds with $r_2 \stackrel{\text{def}}{=} r \restriction (e \setminus \kappa_\xi)$.

For (b) we want that $(r_1, R_1) \Vdash \check{r} \in \bar{\rho}$ which holds because $(\check{r}, (r_1, R_1)) \in \bar{\rho}$.

For (c) we want that

$$(r_1, R_1) \Vdash \forall \pi (\pi \in \rho \to \mathsf{dom}(\pi) = \mathsf{dom}(\check{r}_2) \land (\pi \subseteq \check{r}_2 \lor \pi \supseteq \check{r}_2))$$

or equivalently that

$$\forall \pi \in V^{\mathbb{E}} \forall (b, B) \leq (r_1, R_1) \exists (b', B') \leq (b, B)$$
$$((b', B') \Vdash \neg \pi \in \bar{\rho} \text{ or } (b', B') \Vdash (\mathsf{dom}(\pi) = \mathsf{dom}(\check{r}_2) \land (\pi \subseteq \check{r}_2 \lor \pi \supseteq \check{r}_2))).$$

Let $\pi \in V^{\mathbb{E}}$ and $(b, B) \leq (r_1, R_1)$ be arbitrary and let $(b', B') \leq (b, B)$ decide the formula $\pi \in \bar{\rho}$. Assume that $(b', B') \not\Vdash \neg \pi \in \bar{\rho}$. Then we have that $(b', B') \Vdash \pi \in \bar{\rho}$. By the definition of ρ there is some $r' \in R$ such that

$$(b', B') \Vdash \pi = (r' \restriction (e \setminus \kappa_\xi))^\vee.$$

Since (r, R) is a condition in P_3 we get that

$$(b', B') \Vdash \mathsf{dom}(\pi) = \mathsf{dom}(\check{r}_2) \land (\pi \subseteq \check{r}_2 \lor \pi \supseteq \check{r}_2).$$

For (d) we want to show that for every $\pi \in V^{\mathbb{E}}$,

$$(r_1, R_1) \Vdash (\pi \in \bar{\rho} \land \alpha(\pi) = \mathsf{cf}'\alpha(\pi) \to$$
$$\exists \check{X} (\check{X} \in \check{\Phi}_{\alpha(\pi)} \land \forall \check{\beta} (\beta \in \check{X} \leftrightarrow \pi \cup \{(\check{\alpha}(\pi), \check{m}(\pi), \check{\beta})\} \in \bar{\rho}))).$$

As before, let $\pi \in V^{\mathbb{E}}$ and $(b, B) \leq (r_1, R_1)$ be arbitrary and let $(b', B') \leq (b, B)$ decide the formula $\pi \in \bar{\rho} \land \alpha(\pi) = \mathsf{cf}'\alpha(\pi)$. Assume that

$$(b, B) \not\Vdash \neg(\pi \in \bar{\rho} \land \alpha(\pi) = \mathsf{cf}'\alpha(\pi)).$$

Then $(b, B) \Vdash (\pi \in \bar{\rho} \land \alpha(\pi) = \mathsf{cf}'\alpha(\pi))$.

Let $r' \in R$ be such that $(b', B') \leq (r_1, R_1) \uparrow (r' \restriction (e \cap \kappa_\xi))$ and

$$(b', B') \Vdash \pi = (r' \restriction (e \setminus \kappa_\xi))^\vee.$$

Call $r'_1 \stackrel{\text{def}}{=} r' \restriction (e \cap \kappa_\xi)$ and $r'_2 \stackrel{\text{def}}{=} r' \restriction (e \setminus \kappa_\xi)$.

We want to show that

$$(b', B') \Vdash \exists \check{X}(\check{X} \in \check{\Phi}_{\alpha(\pi)} \wedge \forall \check{\beta}(\beta \in \check{X} \leftrightarrow \pi \cup \{(\check{\alpha}(\pi), \check{m}(\pi), \check{\beta})\} \in \bar{\sigma})).$$

Case 1, if $\alpha(r') = \alpha(r'_2) \geq \kappa_\xi$. Then let

$$X \stackrel{\text{def}}{=} \mathsf{Suc}_R(r') = \{\beta \ ; \ r' \cup \{(\alpha(r'), m(r'), \beta)\} \in R\},$$

and note that $X \in \Phi_{\alpha(r'_2)}$ and

$$\begin{aligned} r' \cup \{(\alpha(r'), m(r'), \beta)\} \in R &\Longleftrightarrow ((r'_2 \cup \{(\alpha(r'_2), m(r'_2), \beta)\})^\vee, (r_1, R_1) \uparrow (r'_1)) \in \bar{\rho} \\ &\Longleftrightarrow \beta \in X. \end{aligned}$$

Let $\check{\beta} \in V^{\mathbb{E}}$ be arbitrary. We want that

$$(b', B') \Vdash (\check{\beta} \in \check{X} \wedge \pi \cup \{(\check{\alpha}(\pi), \check{m}(\pi), \check{\beta})\} \in \bar{\rho}) \vee (\check{\beta} \notin \check{X} \wedge \pi \cup \{(\check{\alpha}(\pi), \check{m}(\pi), \check{\beta})\} \notin \bar{\rho}),$$

or equivalently that

$$\begin{aligned} &\forall (c, C) \leq (b', B') \exists (c', C') \leq (c, C) \\ &\qquad ((c', C') \Vdash (\check{\beta} \in \check{X} \wedge \pi \cup \{(\check{\alpha}(\pi), \check{m}(\pi), \check{\beta})\} \in \bar{\rho}) \text{ or} \\ &\qquad (c', C') \Vdash (\check{\beta} \notin \check{X} \wedge \pi \cup \{(\check{\alpha}(\pi), \check{m}(\pi), \check{\beta})\} \notin \bar{\rho})). \end{aligned}$$

So let $(c, C) \leq (b', B')$ be arbitrary and let $(c', C') \leq (c, C)$ be stronger than $(r_1, R_1) \uparrow (r'_1)$ and decide $\pi \cup \{(\check{\alpha}(\pi), \check{m}(\pi), \check{\beta})\} \in \bar{\rho}$.

Clearly this (c', C') satisfies

$$\begin{aligned} &(c', C') \Vdash (\check{\beta} \in \check{X} \wedge \pi \cup \{(\check{\alpha}(\pi), \check{m}(\pi), \check{\beta})\} \in \bar{\rho}) \text{ or} \\ &(c', C') \Vdash (\check{\beta} \notin \check{X} \wedge \pi \cup \{(\check{\alpha}(\pi), \check{m}(\pi), \check{\beta})\} \notin \bar{\rho}) \end{aligned}$$

and we're done with this case.

Case 2, if $\alpha(r') < \kappa_\xi$, then let $r'' \supseteq r'$ be such that $r'' \restriction (e \setminus \kappa_\xi) = r'_2$ and $\alpha(r'') = \alpha(r'_2) \geq \kappa_\xi$. The rest follows as in case 1.

For (e), (f), and (g) we proceed similarly. qed claim 2

Claim 3. The map i is a dense embedding.

Proof of claim. Let $((t, T), \sigma) \in \mathbb{E} * \dot{\mathbb{Q}}$ be arbitrary. We want to define an $(r, R) \in J$ such that $i((r, R)) \leq ((t, T), \sigma)$. Define $r \stackrel{\text{def}}{=} t \cup s$. By (a) of the definition of $\dot{\mathbb{Q}}$, $r \in P_2$.

If $r' \in P_2$ is such that $r' \subseteq r$ then let $r' \in R$. Above r we define R recursively as follows. Let $r' \in P_2$ be such that $r' \supseteq r$ and $r' \in R$.

- If $\alpha(r') < \kappa_\xi$ then $r' \cup \{(\alpha(r'), m(r'), \beta)\} \in R :\Longleftrightarrow$
$$(r' \cup \{(\alpha(r'), m(r'), \beta)\}) \restriction (e \cap \kappa_\xi) \in T.$$
- If $\alpha(r') \geq \kappa_\xi$ then $r' \cup \{(\alpha(r'), m(r'), \beta)\} \in R :\Longleftrightarrow$ for some $(t', T') \leq (t, T) \uparrow (r' \restriction (e \cap \kappa_\xi))$,
$$(t', T') \Vdash ((r' \restriction (e \setminus \kappa_\xi)) \cup \{(\alpha(r'), m(r'), \beta)\})^\vee \in \bar{\sigma}.$$

Subclaim 1. For every $r' \in R$, call $r'_1 \stackrel{\text{def}}{=} r' \restriction (e \cap \kappa_\xi)$ and $r'_2 \stackrel{\text{def}}{=} r' \restriction (e \setminus \kappa_\xi)$. We have that there is a $(t', T') \leq (t, T) \uparrow (r'_1)$ such that $(t', T') \Vdash \check{r}'_2 \in \bar{\sigma}$.

Proof of subclaim. Since by the definition of R and $\dot{\mathbb{Q}}$ this holds for all $r' \subseteq r$, we'll use induction with base case $r' = r$. For $r' = r$ it holds with $(t', T') = (t, T)$ due to (b) of the definition of $\dot{\mathbb{Q}}$. So assume it holds for r' and let $r \cup \{(\alpha(r'), m(r'), \beta)\} \in R$ be arbitrary.

If $\alpha(r') < \kappa_\xi$ then $\alpha(r') = \alpha(r'_1)$ and by the definition of R we have that
$$(r'_1 \cup \{(\alpha(r'), m(r'), \beta)\}) \in T.$$
By the induction hypothesis we get that for some $(t', T') \leq (t, T) \uparrow (r'_1 \cup \{(\alpha(r'), m(r'), \beta)\})$, $(t', T') \Vdash \check{r}_2 \in \bar{\sigma}$.

If $\alpha(r') \geq \kappa_\xi$ then $\alpha(r') = \alpha(r'_2)$ and by the definition of R we have that for some $(t', T') \leq (t, T) \uparrow (r'_1)$, $(t', T') \Vdash (r'_2 \cup \{(\alpha(r'_2), m(r'_2), \beta)\})^\vee \in \bar{\sigma}$. qed subclaim 1

Subclaim 2. $(r, R) \in J$

Proof of subclaim. To show that $(r, R) \in P_3$ we only need to verify (6) and (7) of the definition of P_3.

For (6), let $r' \in R$ with $r' \supseteq r$ and $\alpha(r')$ of type 1. Call $r'_1 \stackrel{\text{def}}{=} \restriction (e \cap \kappa_\xi)$ and $r'_2 \stackrel{\text{def}}{=} r' \restriction (e \setminus \kappa_\xi)$.

If $\alpha(r') < \kappa_\xi$ then $r' \cup \{(\alpha(r'), m(r'), \beta)\} \in R$ iff $r'_1 \cup \{(\alpha(r'_1), m(r'_1), \beta)\} \in T$. Since $(t, T) \in \mathbb{E}$ we get that
$$\mathsf{Suc}_R(r') = \{\beta\ ;\ r'_1 \cup \{(\alpha(r'_1), m(r'_1), \beta)\} \in T\} \in \Phi_{\alpha(r')}.$$
So assume that $\alpha(r') \geq \kappa_\xi$. We have that
$$(t, T) \Vdash \forall \pi (\pi \in \bar{\sigma} \wedge \pi \supseteq \check{s} \wedge \alpha(\pi) = \mathsf{cf}'\alpha(\pi) \rightarrow$$
$$\exists \check{X} (\check{X} \in \check{\Phi}_{\alpha(\pi)} \wedge \forall \check{\beta} (\check{\beta} \in \check{X} \leftrightarrow \pi \cup \{(\alpha(\pi), m(\pi), \beta)\} \in \bar{\sigma}))).$$
By Subclaim 1 we have that for some $(t', T') \leq (t, T) \uparrow (r'_1) \leq (t, T)$,
$$(t', T') \Vdash \check{r}'_2 \in \bar{\sigma} \wedge \check{r}'_2 \supseteq \check{s} \wedge \alpha(\check{r}'_2) = \mathsf{cf}'\alpha(\check{r}'_2),$$

thus $(t',T') \Vdash \exists \check{X}(\check{X} \in \check{\Phi}_{\alpha(\check{r}'_2)} \wedge \forall \check{\beta}(\check{\beta} \in \check{X} \leftrightarrow \check{r}'_2 \cup \{(\alpha(\check{r}'_2), m(\check{r}'_2), \beta)\} \in \bar{\sigma}))$. So for some $(t'',T'') \leq (t',T')$ there is some $X \in \Phi_{\alpha(r'_2)}$ such that for every $\check{\beta} \in V^{\mathbb{E}}$ we have that

$$(t'',T'') \Vdash \check{\beta} \in \check{X} \leftrightarrow \check{r}'_2 \cup \{(\alpha(\check{r}'_2), m(\check{r}'_2), \beta)\} \in \bar{\sigma}.$$

Let $\beta \in X$. Then $(t'',T'') \Vdash \check{r}'_2 \cup \{(\alpha(\check{r}'_2), m(\check{r}'_2), \beta)\} \in \bar{\sigma}$ which by the definition of R means that $r' \cup \{(\alpha(r'), m(r'), \beta)\} \in R$. So $X \subseteq \mathsf{Suc}_R(r') \in \Phi_{\alpha(r)}$.

For (7), let $r' \in R$ with $r'. \supseteq r$ and $\alpha(r')$ of type 2. Again, call $r'_1 \stackrel{\text{def}}{=} r' \restriction (e \cap \kappa_\xi)$ and $r'_2 \stackrel{\text{def}}{=} r' \restriction (e \setminus \kappa_\xi)$.

If $\alpha(r') \geq \kappa_\xi$ and $\mathsf{cf}'\alpha(r') < \kappa_\xi$ then we have that for every $\pi \in V^{\mathbb{E}}$,

$$\begin{aligned}(t,T) \Vdash &(\pi \in \bar{\sigma} \wedge \pi \supseteq \check{s} \wedge \alpha(\pi) > \mathsf{cf}'\alpha(\pi) < \kappa_\xi \rightarrow \\ &\exists \check{X}(\check{X} \in \check{\Phi}_{\alpha(\pi),\Gamma(\mathsf{cf}'\alpha(\pi))(m(\pi))} \wedge \forall \check{\beta}(\check{\beta} \in \check{X} \leftrightarrow \pi \cup \{(\alpha(\pi), m(\pi), \beta)\} \in \bar{\sigma}))).\end{aligned}$$

By Subclaim 1 we have that for some $(t',T') \leq (t,T) \uparrow (r'_1) \leq (t,T)$,

$$(t',T') \Vdash \check{r}'_2 \in \bar{\sigma} \wedge \check{r}'_2 \supseteq \check{s} \wedge \alpha(\check{r}'_2) > \mathsf{cf}'\alpha(\check{r}'_2) < \kappa_\xi.$$

Therefore for some $(t'',T'') \leq (t',T')$ there is some $X \in V$ such that $(t'',T'') \Vdash \check{X} \in \Phi_{\alpha(\check{r}'_2),\Gamma(\mathsf{cf}'\alpha(\check{r}'_2))(m(\check{r}'_2))}$ and for every $\beta \in V^{\mathbb{E}}$ we have that

$$(t'',T'') \Vdash \check{\beta} \in \check{X} \leftrightarrow \check{r}'_2 \cup \{(\alpha(\check{r}'_2), m(\check{r}'_2), \beta)\} \in \bar{\sigma}.$$

But since $r'_1 \cup r'_2 = r' \in P_2$, we have that $(t,T) \uparrow (r'_1)$ decides the value of $\Gamma(\mathsf{cf}'\alpha(\check{r}'_2))(m(\check{r}'_2))$ to be $\gamma \stackrel{\text{def}}{=} r'_1(\mathsf{cf}'\alpha(r'_2))(m(r'_2))$. So $(t'',T'') \Vdash \check{X} \in \Phi_{\mathsf{cf}'\alpha(\check{r}'_2),\gamma}$. Since $X \in V$ and $\Phi_{\mathsf{cf}'\alpha(\check{r}'_2),\gamma} \in V$ we have that $X \in \Phi_{\mathsf{cf}'\alpha(\check{r}'_2),\gamma}$. So take an arbitrary $\beta \in X$. Then $(t'',T'') \Vdash r'_2 \cup \{(\alpha(\check{r}'_2), m(\check{r}'_2), \beta)\} \in \bar{\sigma}$ which by the definition of R means that $r' \cup \{(\alpha(r'), m(r'), \beta)\} \in R$ and consequently $X \subseteq \mathsf{Suc}_R(r') \in \Phi_{\alpha(r),r(\mathsf{cf}'\alpha(r))(m(r))}$.

Similarly for the other cases where $\alpha(r') > \kappa_\xi$ and $\mathsf{cf}'\alpha(r') \geq \kappa_\xi$, and $\alpha(r') < \kappa_\xi$.

To conclude Subclaim 2 we want to show that the last condition for membership in J is fulfilled, i.e., if $r' \in R$, $\alpha \geq \kappa_\xi$, and $n < \omega$ are such that $(\alpha, n) \in \mathsf{dom}^2(r') \setminus \mathsf{dom}^2(t)$ and $\mathsf{cf}'\alpha < \alpha$, then $\Phi_{\alpha,r'(\mathsf{cf}'\alpha)(n)}$ is κ_ξ-complete. Let $q \subseteq r'$ be such that for some $q' \in R$, $q = q' \cup \{(\alpha, n, \beta)\}$, $\alpha(q') = \alpha$, and $m(q') = n$. Note that $r'(\mathsf{cf}'\alpha)(n) = q'(\mathsf{cf}'\alpha)(n)$.

If $\mathsf{cf}'\alpha \geq \kappa_\xi$ then clearly $\Phi_{\alpha,q'(\mathsf{cf}'\alpha)(n)}$ is κ_ξ-complete.

If $\mathsf{cf}'\alpha < \kappa_\xi$ then note that $q' \restriction (e \cap \kappa_\xi) \in T$ and $(t,T) \in \mathbb{E}$, i.e., for some $(s,S) \in J$, $(s,S) \restriction^* E_e = (t,T)$. So $q'(\mathsf{cf}'\alpha)(n)$ must be high enough for the ultrafilter $\Phi_{\alpha,q'(\mathsf{cf}'\alpha)(n)}$ to be κ_ξ-complete.
qed subclaim 2

Lastly, we want to show that $i((r,R)) \leq_{\mathbb{E} * \dot{\mathbb{Q}}} ((t,T),\sigma)$. Let $i((r,R))) = ((r_1,R_1),\rho)$, and $\rho = \mathsf{op}(\check{r}'_2, \bar{\rho})$. By the definition of R and of i we immediately get that $(r_1,R_1) \leq_{\mathbb{E}} (t,T)$. So it remains to show that $(r_1, R_1) \Vdash \bar{\rho} \subseteq \bar{\sigma}$, i.e.,

$$(r_1, R_1) \Vdash \forall \pi (\pi \in \bar{\rho} \to \pi \in \bar{\sigma}),$$

or equivalently that

$$\forall \pi \in V^{\mathbb{E}} \forall (b, B) \leq (r_1, R_1) \exists (b', B') \leq (b, B)((b', B') \Vdash \neg \pi \in \bar{\rho} \text{ or } (b', B') \Vdash \pi \in \bar{\sigma}).$$

So let $\pi \in V^{\mathbb{E}}$ and $(b, B) \leq (r_1, R_1)$ be arbitrary. There is some $(b', B') \leq (b, B)$ that decides "$\pi \in \bar{\rho}$". Assume that $(b', B') \not\Vdash \neg \pi \in \bar{\rho}$. Then $(b', B') \Vdash \pi \in \bar{\rho}$. By the definition of $\bar{\rho}$, there must be some $r' \in R$ such that $(b', B') \Vdash \pi = (r' \upharpoonright (e \setminus \kappa_\xi))^\vee$. Call $r'_1 \stackrel{\text{def}}{=} r' \upharpoonright (e \cap \kappa_\xi)$ and $r'_2 := r' \upharpoonright (e \setminus \kappa_\xi)$.

By Subclaim 2 we have that there is some $(t', T') \leq (t, T)$ such that $(t', T') \Vdash \check{r}_2 \in \bar{\sigma}$. Since (b', B') was arbitrary above (b, B) such that it decides "$\pi \in \rho$", we can assume it is also stronger than (t', T'). Therefore $(b', B') \Vdash \pi \in \bar{\sigma}$. qed claim 3

So we have shown that $\mathbb{Q}$ can indeed be seen as the top part of E_e, cut in κ_ξ. For the rest of the proof we will not work with the name $\dot{\mathbb{Q}}$ but with $\mathbb{Q}$ inside $V[G^*]$.

Claim 4. In $V[G^*]$, $(\mathbb{Q}, \leq^*_{\mathbb{Q}})$ is κ_ξ-closed.

Proof of claim. Let $\gamma < \kappa_\xi$ and let $\{S_\zeta \ ; \ \zeta < \gamma\} \in V[G^*]$ be a $\leq^*_{\mathbb{Q}}$-descending sequence of elements in $\mathbb{Q}$. Since $E_{e \cap \kappa_\xi}$ is small forcing with respect to κ_ξ, we can use the Lévy-Solovay theorem to get that all ultrafilters involved in the definition of $\mathbb{Q}$ can be extended to κ_ξ-complete ultrafilters in $V[G^*]$. Then for some $\bar{S} \subset \bigcup_{\zeta<\gamma} S_\zeta$, $\bar{S} \in V$, the set $\bar{S}$ is a condition in $\mathbb{Q}$ and it is stronger than all of the S_ζ. qed claim 4

So as usual with Prikry-like forcings, it remains to show the following Prikry-like lemma for $\mathbb{Q}$.

Claim 5. (The Prikry lemma for $\mathbb{Q}$) In $V[G^*]$, let $\tau_1, \ldots, \tau_k$ be $\mathbb{Q}$-names, and φ be a formula with k free variables. Then for every forcing condition $(s, S) \in \mathbb{Q}$ there is a stronger condition $(s, W) \in \mathbb{Q}$ which decides $\varphi(\tau_1, \ldots, \tau_n)$.

This proof is almost identical to Gitik's Prikry style lemma [**Git80**, Lemma 5.1].

Proof of claim. Work in $V[G^*]$. Let $(s, S) \in \mathbb{Q}$.

Let $r \in S$. If $\alpha(r)$ is of type 1 then call

$$\Phi_r \stackrel{\text{def}}{=} \Phi_{\alpha(r)}.$$

If $\alpha(r)$ is of type 2 and $\mathsf{cf}'\alpha(r) \geq \kappa_\xi$ then call

$$\Phi_r \stackrel{\text{def}}{=} \Phi_{\alpha(r),r(\mathsf{cf}'\alpha(r))(m(r))}.$$

If $\alpha(r)$ is of type 2 and $\mathsf{cf}'\alpha(r) < \kappa_\xi$ then let $\gamma \in \mathsf{cf}'\alpha$ be such that $\bigcup G^*(\mathsf{cf}'\alpha(r))(m(r)) = \gamma$, and call

$$\Phi_r \stackrel{\text{def}}{=} \Phi_{\alpha(r),\gamma}.$$

For all $r \in P_2 \cap (e \setminus \kappa_\xi) \times \omega \times \kappa_\rho$ define

$$\bar{\Phi}_r \stackrel{\text{def}}{=} \{X \subseteq \alpha(r)\ ;\ \exists Y \in \Phi_r(Y \subseteq X)\}.$$

Let $\theta < \kappa_\xi$ be a cardinal of V such that $E_{e\cap\kappa_\xi}$ has the θ-cc in V. For each $r \in S$, the ultrafilter Φ_r is at least κ_ξ complete. So we can use arguments from the Lévy-Solovay theorem to get that in $V[G^*]$, each $\bar{\Phi}_r$ is at least κ_ξ-complete as well. Also define the following sets.

$$\begin{aligned}
S_0 &\stackrel{\text{def}}{=} \{r \in S\ ;\ r \supseteq s \text{ and } \exists R \subseteq S((s,R) \in \mathbb{Q} \text{ and } (s,R) \Vdash_{\mathbb{Q}} \varphi(\tau_1,\ldots,\tau_k))\}\\
S_1 &\stackrel{\text{def}}{=} \{r \in S\ ;\ r \supseteq s \text{ and } \exists R \subseteq S((s,R) \in \mathbb{Q} \text{ and } (s,R) \Vdash_{\mathbb{Q}} \neg\varphi(\tau_1,\ldots,\tau_k))\}\\
S_2 &\stackrel{\text{def}}{=} \{r \in S\ ;\ r \supseteq s \text{ and } \forall R \subseteq S(\text{if } (s,R) \in \mathbb{Q} \text{ then } (s,R) \text{ does not decide}\\
&\qquad \varphi(\tau_1,\ldots,\tau_k))\}.
\end{aligned}$$

Clearly, $S = S_0 \cup S_1 \cup S_2$. Let $e \setminus \kappa_\xi \stackrel{\text{def}}{=} \{\alpha_0,\ldots,\alpha_{n-1}\}$. We will now enumerate the set $((e \setminus \kappa_\xi) \times \omega) \setminus \mathsf{dom}^2(s)$ from left to right and upwards, by a function x that is recursively defined as follows. First,

$$x(0) \stackrel{\text{def}}{=} (\alpha(s), m(s)).$$

Now let $x(i) = (\alpha_j, m)$ for some $\alpha_j \in e \setminus \kappa_\xi$ and $m \in \omega$. If $j < n-1$ then let

$$x(i+1) \stackrel{\text{def}}{=} (\alpha_{j+1}, m),$$

and if $j = n-1$ then

$$x(i+1) \stackrel{\text{def}}{=} (\alpha_0, m+1).$$

For every $i \in \omega$ define

$$\tilde{F}_i \stackrel{\text{def}}{=} \{r \in S\ ;\ \mathsf{dom}^2(r) \setminus \mathsf{dom}^2(s) = x``(i+1)\}.$$

We could say that this is the set of $r \in S$ whose x-distance from s is i. Now for $i \leq j$ we will define recursively on $i - j$ a set of functions $F_{i,j} : \tilde{F}_i \to 3$. For $\ell < 3$ and $r \in \tilde{F}_i$ let

$$F_{i,i}(r) = \ell :\Longleftrightarrow r \in S_\ell.$$

For $i < j$ let $F_{i,j}(r) = \ell :\Longleftrightarrow$ the set

$$\{\beta\ ;\ r \cup \{(\alpha(r), m(r), \beta)\} \in S \text{ and } F_{i+1,j}(r \cup \{(\alpha(r), m(r), \beta)\}) = \ell\}$$

is in the ultrafilter Φ_r. Define recursively on $i < \omega$ a subset $\tilde{F}'_i \subset \tilde{F}_i$. Using the definition and the ω-completeness of the ultrafilter $\bar{\Phi}_s$, we find (in V), a set $\tilde{F}'_0 \subseteq \tilde{F}_0$ which is homogeneous for all functions in the set $\{F_{0,j}\ ;\ j<\omega\}$ and which is such that the set

$$\{\beta\ ;\ s\cup\{(\alpha(s),m(s),\beta)\}\in\tilde{F}'_0\}$$

is in Φ_s. By *homogeneous* here we mean that for all $t_1,t_2\in\tilde{F}'_0$ and for all $0\le j<\omega$, $F_{0,j}(t_1)=F_{0,j}(t_2)$. For $i>0$ we take

$$\tilde{F}'_i \stackrel{\text{def}}{=} \{r\in\tilde{F}_i\ ;\ r^-\in\tilde{F}'_{i-1} \text{ and } \forall i\le j(F_{i,j}(r)=F_{i-1,j}(r^-))\},$$

where r^- is defined as in Definition 2.35(7). By the induction hypothesis, it follows that $\tilde{F}'_i$ is homogeneous for all functions in the set $\{F_{i,j}\ ;\ i\le j<\omega\}$. The definition of the functions $F_{i,j}$ implies that for every $r\in\tilde{F}'_{i-1}$,

$$\{\beta\ ;\ r\cup\{(\alpha(r),m(r),\beta)\}\in\tilde{F}'_i\}\in\Phi_r. \tag{3}$$

Define the set

$$\tilde{F}\stackrel{\text{def}}{=}\{s\}\cup\bigcup\{\tilde{F}'_i\ ;\ i<\omega\}.$$

Let's see whether this leads us in the right direction, we will show the following.

Subclaim. If $s_1,s_2\in\tilde{F}$, $(s_1,A_1),(s_2,A_2)\in\mathbb{Q}$, and $(s_1,A_1),(s_2,A_2)\le_{\mathbb{Q}}(s,S)$, then it is impossible to have that $(s_1,A_1)\Vdash_{\mathbb{Q}}\varphi(\tau_1,\dots,\tau_k)$ and $(s_2,A_2)\Vdash_{\mathbb{Q}}\neg\varphi(\tau_1,\dots,\tau_k)$.

Proof of subclaim. We have that for some $i_1,i_2<\omega$ and for every $j=1,2$,

$$\mathsf{dom}^2(s_j)=\mathsf{dom}^2(s)\cup\{x(\ell)\ ;\ \ell<i_j\}.$$

Without loss of generality we may assume that $i_1\le i_2$. If $i_1<i_2$ then we can increase the $\mathsf{dom}^2(s_1)$, one step at a time until we get $i_1=i_2$. We have that the set

$$E\stackrel{\text{def}}{=}\{\beta<\alpha(s_1)\ ;\ s_1\cup\{(\alpha(s_1),m(s_1),\beta)\}\in\tilde{F}'_{i_1}\} \text{ is in } \Phi_{s_1}$$

and since (s_1,A_1) is a condition in $\mathbb{Q}$ we have that also the set

$$E'\stackrel{\text{def}}{=}\{\beta<\alpha(s_1)\ ;\ s_1\cup\{(\alpha(s_1),m(s_1),\beta)\}\in A_1\} \text{ is in } \Phi_{s_1}.$$

Let $\beta\in E\cap E'$, let $\bar{s}_1\stackrel{\text{def}}{=}s_1\cup\{(\alpha(s_1),m(s_1),\beta)\}$, and let $\bar{A}_1\stackrel{\text{def}}{=}\{t\in A_1\ ;\ t\supseteq\bar{s}_1\}$. Then we have that $\bar{s}_1\in\tilde{F}'_{i_1}\subseteq\tilde{F}$ and that $(\bar{s}_1,\bar{A}_1)\le_{\mathbb{Q}}(s_1,A_1)$. Therefore, $(\bar{s}_1,\bar{A}_1)\Vdash_{\mathbb{Q}}\varphi(\tau_1,\dots,\tau_n)$, and

$$\mathsf{dom}^2(\bar{s}_1)=\mathsf{dom}^2(s)\cup\{x(k)\ ;\ k<i_1+1\}.$$

This way we keep increasing i_1 until we get $i_1=i_2$. Denote i_1-i_2 by i. If $i=0$ then $s_1=s_2=r$, therefore (s_1,A_1) and (s_2,A_2) are compatible which is a contradiction. If $i>0$ then $s_1,s_2\in\tilde{F}'_{i-1}$. Because $(s_1,A_1)\le_{\mathbb{Q}}(s,S)$ and $(s_1,A_1)\Vdash_{\mathbb{Q}}\varphi(\tau_1,\dots,\tau_n)$, we have that $s_1\in S_0$, thus $F_{i-1,i-1}(s_1)=0$. Similarly we get that $F_{i-1,i-1}(s_2)=1$ which contradicts the homogeneity of $\tilde{F}'_{i-1}$ for $F_{i-1,i-1}$. qed subclaim

So to finish the proof of this claim, we first show that $(s,\tilde{F})$ is indeed a condition in $\mathbb{Q}$. It suffices to show that for every $s'\in\tilde{F}$, the set $\mathsf{Suc}_{\tilde{F}}(s')$ of successors of s' in $\tilde{F}$ is in the ultrafilter $\Phi_{s'}$. We have that

$$\begin{aligned}\mathsf{Suc}_{\tilde{F}}(s') =& \{\beta < \alpha(s') \ ; \ s' \cup \{(\alpha(s'), m(s'), \beta)\} \in \tilde{F}\} \\ & \{\beta < \alpha(s') \ ; \ s' \cup \{(\alpha(s'), m(s'), \beta)\} \in \tilde{F}_{i+1}, \}\end{aligned}$$

where $i \in \omega$ is such that $s' \in \tilde{F}_i$. By 3 we get that $\mathsf{Suc}_{\tilde{F}}(s') \in \Phi_{s'}$.

Finally, let $(s_1, A_1) \in \mathbb{Q}$ be any condition that decides $\varphi(\tau_1, \ldots, \tau_n)$. Without loss of generality assume that $(s_1, A_1) \Vdash_{\mathbb{Q}} \varphi(\tau_1, \ldots, \tau_n)$. By the approximation lemma we can assume that $\mathsf{dom}(s_1) = e$, and hence $s_1 \in \tilde{F}$ and $A_1 \subseteq \tilde{F}$. Suppose that $(s, \tilde{F})$ does not decide $\varphi(\tau_1, \ldots, \tau_n)$, then there must be a $(s_2, A_2) \leq_{\mathbb{Q}} (s, \tilde{F})$ such that $(s_2, A_2) \Vdash_{\mathbb{Q}} \neg\varphi(\tau_1, \ldots, \tau_n)$. But this contradicts the subclaim. Therefore $(s, \tilde{F})$ decides $\varphi(\tau_1, \ldots, \tau_n)$. qed claim 5

So with the standard Prikry style arguments we can see that $\mathbb{Q}$ does not add bounded subsets to κ_ξ, therefore E_e cannot collapse κ_ξ, and so a function like f in the beginning of the proof cannot exist in $V[G \restriction^* E_e]$. qed

Next we will see that we singularised the targeted ordinals. This is similar to [**Git80**, Lemma 3.4].

LEMMA 2.38. *Every cardinal in* $(\mathsf{Reg}^{\kappa_\rho})^V$ *has cofinality* ω *in* $V(G)$. *Thus every cardinal in the interval* (ω, κ_ρ) *is singular.*

PROOF. Let $\alpha \in \mathsf{Reg}^{\kappa_\rho}$. For every $\beta < \alpha$, the set

$$D_\beta := \{(t, T) \in P_3 \ ; \ \exists n < \omega (t(\alpha)(n) \geq \beta)\}$$

is dense in $(P_3, \leq)$. Hence $f_\alpha := \bigcup\{t(\alpha) \ ; \ (t, T) \in G\}$ is a function from ω onto an unbounded subset of α. This function has a symmetric name, which is supported by $E_{\{\alpha\}}$. Therefore $f_\alpha \in V(G)$. qed

Now we will show that in the interval (ω, κ_ρ), *only* the former strongly compact cardinals and their (singular) limits remain cardinals, i.e., that all cardinals of V that are between the strongly compacts and their (singular) limits have collapsed.

LEMMA 2.39. *For every ordinal* $\xi \in [-1, \rho)$ *and every* $\alpha \in (\kappa_\xi, \kappa_{\xi+1})$, $(|\alpha| = \kappa_\xi)^{V(G)}$.

PROOF. Fix an ordinal $\xi \in [-1, \rho)$. Since strongly compact cardinals are limits of inaccessible cardinals, it suffices to show that for every inaccessible $\alpha \in (\kappa_\xi, \kappa_{\xi+1})$, we have that $(|\alpha| = |\kappa_\xi|)^{V(G)}$.

Let $\alpha \in (\kappa_\xi, \kappa_{\xi+1})$ be inaccessible. We will use the bijection h_α to show that $E_{\{\alpha\}}$ is isomorphic to the strongly compact injective tree-Prikry forcing $\mathbb{P}^{\mathsf{st}}_{U_\alpha}$ with respect to the ultrafilter U_α.

Towards the isomorphism, define a function f from the injective finite sequences of elements of $\mathcal{P}_{\kappa_\xi}(\alpha)$ to P_2 by

$$f(t) \stackrel{\text{def}}{=} \{(\alpha, m, \beta) \; ; \; m \in \mathsf{dom}(t) \wedge \beta = h_\alpha(t(m))\}.$$

Define another function $i : \mathbb{P}^{\mathsf{st}}_{U_\alpha} \to E_{\{\alpha\}}$ by

$$i(T) \stackrel{\text{def}}{=} (f(\mathsf{tr}_T), \{f(t) \; ; \; t \in T \wedge t \trianglerighteq \mathsf{tr}_T\}).$$

This i is indeed a function from T to $E_{\{\alpha\}}$ because h_α is a bijection. In fact, this i is a bijection itself. It is easy to see that it also preserves the $\leq$ relation of the forcings, so $\mathbb{P}^{\mathsf{st}}_{H_\alpha}$ and $E_{\{\alpha\}}$ are isomorphic.

Therefore in any forcing extension of V via P^s_α, α has become a countable union of sets of cardinality less than κ_ξ and therefore is collapsed to κ_ξ. So there is an $E_{\{\alpha\}}$-name for a collapsing function from κ_ξ to α, which can be seen as a P_3-name in HS for such a function, supported by $E_{\{\alpha\}}$. qed

Next we show that the regular cardinals of type 2 have collapsed to the singular limit of strongly compacts below them.

LEMMA 2.40. *For every α of type 2, if β is the largest limit of strongly compacts below α, then $(|\alpha| = \beta)^{V(G)}$.*

PROOF. This proof is very similar to the proof of Lemma 2.28. Similarly to the proof of the previous lemma we assume that α is inaccessible and we look at each of the bijections $h_{\alpha,\nu} : \mathcal{P}_{\kappa^\alpha_\nu}(\alpha) \to \alpha$. Let e be the smallest finite subset of $\mathsf{Reg}^{\kappa_\rho}$ that contains α and is closed under cf'. Look at $V[G \restriction^* E_e]$. Let $\langle \gamma_i \; ; \; i \in \omega \rangle$ be the Prikry sequence added to $\mathsf{cf}'\alpha$, and let $\langle \alpha_i \; ; \; i \in \omega \rangle$ be the Prikry sequence added to α. For each $i \in \omega$, let

$$A_i \stackrel{\text{def}}{=} h^{-1}_{\alpha,\gamma_i}(\alpha_i).$$

We want to show that for each $\delta \in \alpha$, there is some $i \in \omega$ such that $\delta \in A_i$. Fix $\delta \in \alpha$. For all $i \in \omega$, the V-ultrafilter H_{α,γ_i} is fine, so

$$\{A \in \mathcal{P}_{\kappa^\alpha_{\gamma_i}}(\alpha) \; ; \; \delta \in A\} \in H_{\alpha,\gamma_i}.$$

So for every $i \in \omega$, the set

$$Z_i \stackrel{\text{def}}{=} \{\zeta \in \alpha \; ; \; \delta \in h^{-1}_{\alpha,\gamma_i}(\zeta)\} \in \Phi_{\alpha,\gamma_i}.$$

Define the set

$$D_\delta \stackrel{\text{def}}{=} \{(t, T) \in E_e \; ; \; \exists i \in \mathsf{dom}(t)(\delta \in h^{-1}_{\alpha,\gamma_i}(t(\alpha)(i)))\}.$$

This is dense in E_e and δ was arbitrary. Therefore in $V[G \restriction^* E_e]$, we have that $\alpha = \bigcup_{i \in \omega} A_i$ is a countable union of $\leq \beta$-sized sets, and thus there is a symmetric name for a collapse of α to β, supported by E_e. qed

We summarise our results on the cardinal structure of the interval (ω, κ_ρ).

COROLLARY 2.41. *An uncountable cardinal of $V(G)$ that is less than or equal to κ_ρ is a successor cardinal in $V(G)$ iff it is in $\{\kappa_\xi\ ;\ \xi \leq \rho\}$. Thus in $V(G)$, for every $\xi \leq \rho$ we have that $\kappa_\xi = \aleph_{\xi+1}$.*

Also, an uncountable cardinal of $V(G)$ that is less than or equal to κ_ρ is a limit cardinal in $V(G)$ iff it is a limit in the sequence $\langle \kappa_\alpha\ ;\ \alpha < \rho \rangle$ in V.

PROOF. This follows inductively, using Theorem 2.37, the proof of Lemma 2.38, and Lemma 2.39 and Lemma 2.40. qed

Before we go into the combinatorial properties in $V(G)$, let us mention that the Axiom of Choice fails really badly in this model. The following is [**Git80**, Theorem 6.3].

LEMMA 2.42. *In $V(G)$, countable unions of countable sets are not necessarily countable. In particular, every set in H_{κ_ρ} is a countable union of sets of smaller cardinality. Here "x has a smaller cardinality than y" means that x is a subset of y and there is no bijection between them.*

Therefore, $\mathsf{AC}_\omega(\mathcal{P}(\omega))$ fails in this model (also just because ω_1 is singular).

5.2. Results. After we established the cardinal pattern in $V(G)$ it is fairly straightforward to see that we have the following.

LEMMA 2.43. *In $V(G)$, for each $1 \leq \beta \leq \rho$, $\aleph_\beta$ is singular and $\aleph_{\rho+1}$ is a measurable cardinal carrying a normal measure.*

This is because as we saw in Chapter 1, Section 3, measurable cardinals are preserved under small symmetric forcing, which is what this forcing is with respect to κ_ρ. Therefore we can also get the same result with "measurable" replaced by "weakly compact", "Ramsey", a partition property, etc..

The construction in this section is a generalised construction. For particular results, e.g., $\aleph_{\omega+3}$ becoming both the first uncountable regular cardinal and a measurable cardinal, we just put $\rho = \omega + 2$. Thus we can immediately get theorems such as the following.

LEMMA 2.44. *If V is a model of "There is an $\omega+2$-sequence of strongly compact cardinals with a measurable cardinal above this sequence", then there is a symmetric model in which $\aleph_{\omega+3}$ is both a measurable cardinal and the first regular cardinal.*

Just like in Lemma 2.20 we can show that we have the following combinatorial residue from the strongly compacts.

LEMMA 2.45. *In $V(G)$ every cardinal in $(\omega, \kappa_\rho]$ is an almost Ramsey cardinal.*

And just like in Lemma 2.31 we can show the following.

LEMMA 2.46. *In $V(G)$ every limit cardinal in $(\omega, \kappa_\rho]$ is a Rowbottom cardinal carrying a Rowbottom filter.*

We also get the same corollary to this lemma.

COROLLARY 2.47. *If we assumed that in our ground model for every* $\xi \in (0,\rho)$, κ_ξ *is moreover a limit of measurable cardinals, then in* $V(G)$ *we would have that every cardinal in* (ω, κ_ρ) *is a Rowbottom cardinal carrying a Rowbottom filter.*

We conjecture that a modification of the constructions in this chapter could give entire cardinal intervals with successive Rowbottom cardinals carrying Rowbottom filters, without increasing the consistency strength of the assumptions.

3

Chang conjectures and indiscernibles

1. Facts and definitions

This chapter is a consistency strength analysis between model theoretic and combinatorial principles. We will base our study on Chang conjectures and take a look at some neighbouring principles as well.

We start with the definitions of the principles that we will be looking at, and some basic facts about them.

1.1. Chang conjectures, Erdős cardinals, and indiscernibles.

DEFINITION 3.1. For infinite cardinals $\kappa_0 < \kappa_1 < \cdots < \kappa_n$ and $\lambda_0 < \lambda_1 < \cdots < \lambda_n$, a Chang conjecture is the statement

$$(\kappa_n, \ldots, \kappa_0) \twoheadrightarrow (\lambda_n, \ldots, \lambda_0),$$

which we define to mean that for every first order structure

$$\mathcal{A} = \langle \kappa_n, f_i, R_j, c_k \rangle_{i,j,k \in \omega}$$

with a countable language there is an elementary substructure $\mathcal{B} \prec \mathcal{A}$ of cardinality λ_n such that for every $i \leq n$, $|\mathcal{B} \cap \kappa_i| = \lambda_i$.

Since the structures we will consider will always be wellorderable, we will implicitly assume that the have complete sets of Skolem functions. Thus we will always be able to take Skolem hulls.

We will talk about Chang conjectures where κ_n is a successor cardinal, and we will first discuss Chang conjectures that involve four cardinals, like the original Chang conjecture,

$$(\omega_2, \omega_1) \twoheadrightarrow (\omega_1, \omega).$$

According to Vaught [**Vau63**] this model theoretic relation between cardinals was first considered by Chang.

There has been extensive research on Chang conjectures under AC. As we see in [**LMS90**, 1.8(1)], Silver proved in unpublished work that if the ω_1-Erdős cardinal exists then we can force the original Chang conjecture to be true. Soon afterwards Kunen in [**Kun78**] showed that for every $n \in \omega$, $n \geq 1$, the consistency of the Chang conjecture

$$(\omega_{n+2}, \omega_{n+1}) \twoheadrightarrow (\omega_{n+1}, \omega_n)$$

follows from the consistency of the existence of a huge[1] cardinal. The next year Donder, Jensen, and Koppelberg in [**DJK79**] showed that if the original Chang conjecture is true, then the ω_1-Erdős cardinal exists in an inner model. According to [**LMS90**], the same proof shows that for any infinite cardinals κ, λ, the Chang conjecture

$$(\kappa^+, \kappa) \twoheadrightarrow (\lambda^+, \lambda)$$

implies that there is an inner model in which the μ-Erdős cardinal exists, where $\mu = (\lambda^+)^V$. According to the same source, for many other regular cardinals κ, a Chang conjecture of the form

$$(\kappa^+, \kappa) \twoheadrightarrow (\omega_1, \omega)$$

is equiconsistent with the existence of the ω_1-Erdős cardinal [**LMS90**, 1.10]. Donder and Koepke showed in [**DK83**] that for $\kappa \geq \omega_1$,

$$(\kappa^{++}, \kappa^+) \twoheadrightarrow (\kappa^+, \kappa),$$

then $0^\dagger$ exists, which implies that there is an inner model with a measurable cardinal. A year later Levinski published [**Lev84**] in which paper the existence of $0^\dagger$ is derived from each of the following Chang conjectures:

- for any infinite κ and any $\lambda \geq \omega_1$, the Chang conjecture $(\kappa^+, \kappa) \twoheadrightarrow (\lambda^+, \lambda)$
- for any natural number $m > 1$ and any infinite κ, λ the Chang conjecture $(\kappa^{+m}, \kappa) \twoheadrightarrow (\lambda^{+m}, \lambda)$, and
- for any singular cardinal κ, the Chang conjecture $(\kappa^+, \kappa) \twoheadrightarrow (\omega_1, \omega)$.

In 1988, Koepke improved on some of these results by deriving the existence of inner models with sequences of measurable cardinals [**Koe88**] from Chang conjectures of the form $(\kappa^{++}, \kappa^+) \twoheadrightarrow (\kappa^+, \kappa)$ for $\kappa \geq \omega_1$. Finally, under the axiom of choice we may also get inconsistency from certain Chang conjectures. As we see in [**LMS90**, 1.6], finite gaps cannot be

[1]The definition of a huge cardinal can be found in [**Kan03**, page 331].

increased; for example

$$(\omega_5, \omega_4) \twoheadrightarrow (\omega_3, \omega_1)$$

is inconsistent.

If we remove AC from our assumptions this picture changes drastically. In Section 3 we will get successor cardinals with Erdős-like properties using symmetric forcing, so all sorts of Chang conjectures will become 'accessible'.

The connection between Erdős cardinals and structures with certain elementary substructures lies in the existence of certain sets of indiscernibles.

DEFINITION 3.2. For a structure $\mathcal{A} = \langle A, \ldots \rangle$, with $A \subseteq \mathrm{Ord}$, a set $I \subseteq A$ is a set of indiscernibles if for every $n \in \omega$, every n-ary formula φ in the language for $\mathcal{A}$, and every $\alpha_1, \ldots, \alpha_n, \alpha'_1, \ldots, \alpha'_n$ in I, if $\alpha_1 < \cdots < \alpha_n$ and $\alpha'_1 < \cdots < \alpha'_n$ then

$$\mathcal{A} \models \varphi(\alpha_1, \ldots, \alpha_n) \text{ iff } \mathcal{A} \models \varphi(\alpha'_1, \ldots, \alpha'_n).$$

The set I is called a set of good indiscernibles iff it is as above and moreover we allow parameters that lie below $\min\{\alpha_1, \ldots, \alpha_n, \alpha'_1, \ldots, \alpha'_n\}$, i.e., if moreover for every $x_1, \ldots, x_m \in A$ such that $x_1, \ldots, x_m \leq \min\{\alpha_1, \ldots, \alpha_n, \alpha'_1, \ldots, \alpha'_n\}$, and every $(n+m)$-ary formula φ,

$$\mathcal{A} \models \varphi(x_1, \ldots, x_m, \alpha_1, \ldots, \alpha_n) \text{ iff } \mathcal{A} \models \varphi(x_1, \ldots, x_m, \alpha'_1, \ldots, \alpha'_n).$$

The existence of an Erdős cardinal implies all sorts of four cardinal Chang conjectures. First we will need to get indiscernibles from our Erdős cardinals.

LEMMA 3.3. (ZF) *Assume $\kappa \to (\alpha)_2^{<\omega}$, where α is a limit ordinal. Then for any first order structure $\mathcal{A} = \langle A, \ldots \rangle$, with a countable language, and $\kappa \subseteq A$, there is a set $X \subseteq \kappa$, $\mathrm{ot}(X) \geq \alpha$ of good indiscernibles for $\mathcal{A}$.*

This is [**AK08**, Proposition 8]. As we will see in the next lemma, it is the minimality of an Erdős cardinal that makes the indiscernibles lie high enough for us to get Chang conjectures.

LEMMA 3.4. (ZF) *If λ is a cardinal and $\kappa(\lambda)$ exists then for all infinite $\theta < \kappa(\lambda)$ and $\rho < \lambda, \theta$, the Chang conjecture $(\kappa(\lambda), \theta) \twoheadrightarrow (\lambda, \rho)$ holds.*

PROOF. Let $\kappa = \kappa(\lambda)$ and $\mathcal{A} = \langle \kappa, \ldots \rangle$ be an arbitrary first order structure with a countable language. We want to get a set of good indiscernibles for this structure and then get their $\mathcal{A}$-Skolem hull to get the substructure we want. We can do this because even though we cannot use the axiom of choice, the structure $\mathcal{A}$ is wellorderable. In order for the substructure to be of the right type we need a set of good indiscernibles that lies above θ. To ensure this we will use the minimality of $\kappa(\lambda)$.

Let $g : [\theta]^{<\omega} \to 2$ be a function that doesn't have any homogeneous sets of ordertype λ. Consider the structure

$$\bar{\mathcal{A}} \stackrel{\text{def}}{=} \mathcal{A}^\frown \langle \theta, g{\restriction}[\theta]^n \rangle_{n \in \omega},$$

where θ and each $g\restriction[\theta]^n$ are considered as relations. By Lemma 3.3 there is a set I of good indiscernibles of ordertype λ for this structure. There must be at least one $x \in I \setminus \theta$ otherwise I would be a homogeneous set for g of ordertype λ. By indiscernibility every element of I is above θ.

Let $\mathsf{Hull}(I \cup \rho)$ be the $\mathcal{A}$-Skolem hull of $I \cup \rho$. By Lemma 0.23 we have that

$$|\mathsf{Hull}(I \cup \rho)| \leq |I \cup \rho| + |L| = \lambda.$$

But $\lambda = |I \cup \rho| \leq |\mathsf{Hull}(I \cup \rho)|$, thus $\mathsf{Hull}(I \cup \rho)$ has cardinality λ. Because all the indiscernibles lie above θ and because they are good indiscernibles, they are indiscernibles with respect to parameters below θ. So

$$\rho \leq |\mathsf{Hull}(I \cup \rho) \cap \theta| \leq \omega \cdot \rho = \rho.$$

So the substructure $\mathsf{Hull}(I \cup \rho) \prec \mathcal{A}$ is as we wanted. qed

At this point we should note that Chang conjectures do not imply that some cardinal is Erdős. Before we look at the example of the four cardinal Chang conjecture, let us prove that Chang conjectures are preserved under c.c.c.-forcing.

PROPOSITION 3.5. *Let V be a model of* ZFC *in which for the cardinals $\kappa, \theta, \lambda, \rho$, the Chang conjecture $(\kappa, \lambda) \twoheadrightarrow (\lambda, \rho)$ holds. Assume also that $\mathbb{P}$ is a c.c.c.-forcing. If G is a $\mathbb{P}$-generic filter, then $(\kappa, \lambda) \twoheadrightarrow (\lambda, \rho)$ holds in $V[G]$ as well.*

PROOF. Let $\mathcal{A} \stackrel{\text{def}}{=} \langle \kappa, f_i, R_j, c_k \rangle_{i,j,k \in \omega} \in V[G]$ be arbitrary. Since the language of $\mathcal{A}$ is countable, let $\{\exists x \varphi_n(x) \; ; \; n \in \omega\}$ enumerate the existential formulas of $\mathcal{A}$'s language in a way such that for every $n \in \omega$, the arity $\mathsf{ar}_{\varphi_n} \stackrel{\text{def}}{=} k_n$ is less that n. For every $n \in \omega$ let g_n be the Skolem function that corresponds to φ_n, and let $\dot{g}_n$ be a nice name for g_n as a subset of κ^{k_n}. Since $\dot{g}_n$ is a nice name, it is of the form

$$\dot{g}_n \stackrel{\text{def}}{=} \bigcup \{ \{\check{x}\} \times A_x \; ; \; x \in \kappa^{k_n} \}.$$

Where each A_x is an antichain of $\mathbb{P}$ and since $\mathbb{P}$ has the c.c.c., each A_x is countable. For each $x \in \kappa^{k_n}$, let $A_x \stackrel{\text{def}}{=} \{p_{x,0}, p_{x,1}, p_{x,2}, \ldots\}$. In V define for each $n \in \omega$ a function $g_n : \kappa^{k_n - 1} \times \omega \to \kappa$ as follows:

$$g_n(\alpha_1, \ldots, \alpha_{k_n - 1}, \ell) \stackrel{\text{def}}{=} \begin{cases} \beta & \text{if } p_{\{\alpha_1, \ldots, \alpha_{k_n-1}, \beta\}, \ell} \Vdash \dot{g}_n(\check{\alpha}_1, \ldots, \check{\alpha}_{k_n - 1}) = \check{\beta} \\ 0 & \text{otherwise.} \end{cases}$$

In V consider the structure $\mathcal{C} \stackrel{\text{def}}{=} \langle \kappa, g_n \rangle_{n \in \omega}$. Using the Chang conjecture in V take a Chang substructure $\langle B, g_n \rangle_{n \in \omega} \prec \mathcal{C}$. But then in $V[G]$ we have that $\mathcal{B} \stackrel{\text{def}}{=} \langle B, f_i, R_j, c_k \rangle_{i,j,k \in \omega} \prec \mathcal{A}$ is the elementary substructure we were looking for. qed

LEMMA 3.6. *Let $\kappa, \theta, \lambda, \rho$ be infinite cardinals in a model V of* ZFC, *such that $\kappa > \lambda, \theta$ and $\lambda, \theta > \rho$, and assume that $(\kappa, \theta) \twoheadrightarrow (\lambda, \rho)$. Then there is a generic extension where $(\kappa, \theta) \twoheadrightarrow (\lambda, \rho)$ holds and κ is not the λ-Erdős.*

PROOF. If κ is not the λ-Erdős in V then we are done. So assume that κ is $\kappa(\lambda)$ in V. Let $\mu \geq \kappa$ and consider the partial order $\mathsf{Fn}(\mu \times \omega, 2)$ that adds μ many Cohen reals. This partial order has the c.c.c. so all cardinals are preserved by this forcing and by Proposition 3.5, the Chang conjecture is preserved as well. Now let G be a $\mathsf{Fn}(\mu \times \omega, 2)$-generic filter. We have that $(2^\omega)^{V[G]} \geq \mu > \kappa$. We will show that $\kappa \not\to (\omega_1)^2_2$ so κ is not ξ-Erdős for any $\xi \geq> \omega_1$.

Let $\mathbb{R}$ denote the set of reals and let $g : \kappa \to \mathbb{R}$ be injective. Define $F : [\kappa]^2 \to 2$ by

$$F(\{\alpha, \beta\}) \stackrel{\text{def}}{=} \begin{cases} 1 & \text{if } g(\alpha) <_{\mathbb{R}} g(\beta) \\ 0 & \text{otherwise} \end{cases}$$

If there was an ω_1-sized homogeneous set for F then $\mathbb{R}$ would have an ω_1-long strictly monotonous $<_{\mathbb{R}}$-chain which is a contradiction. qed

By looking at the proof of Lemma 3.4 we recognise that the combinatorial property we really need is not an Erdős cardinal but the following.

DEFINITION 3.7. For cardinals $\kappa > \lambda$ and ordinal $\theta < \kappa$ we define the partition property

$$\kappa \to^\theta (\lambda)^{<\omega}_2$$

to mean that for every first order structure $\mathcal{A} = \langle \kappa, \ldots \rangle$ with a countable language, there is a set $I \in [\kappa \setminus \theta]^\lambda$ of good indiscernibles for $\mathcal{A}$. We call such a κ an Erdős-like cardinal with respect to θ, λ.

Note that for any cardinal $\mu > \kappa(\lambda)$ and any $\theta < \kappa(\lambda)$, we have that $\mu \to^\theta (\lambda)^{<\omega}_\rho$. So the existence of cardinals κ, λ, θ such that $\kappa \to^\theta (\lambda)^{<\omega}_2$ is a trivial consequence of the existence of $\kappa(\lambda)$.

COROLLARY 3.8. (ZF) *For κ, θ, λ infinite cardinals with $\kappa > \theta, \lambda$, if $\kappa \to^\theta (\lambda)^{<\omega}_2$ then for every $\rho < \lambda, \theta$ we have $(\kappa, \theta) \twoheadrightarrow (\lambda, \rho)$.*

The property $\kappa \to^\theta (\lambda)^{<\omega}_2$ implies that the Erdős cardinal $\kappa(\lambda)$ exists, and it is much easier to use since it does not require the minimality of the cardinal κ. At this point one could consider cardinals μ, ν that satisfy the properties

$$\mu \to^{\kappa(\lambda)} (\lambda)^{<\omega}_2,$$

$$\nu \to^{\mu} (\lambda)^{<\omega}_2, \text{ etc..}$$

An immediate observation about these cardinal properties is that, e.g., $\mu \to^{\kappa(\lambda)} (\lambda)^{<\omega}_2$ implies that $\kappa(\lambda)$ exists, and the existence of $\kappa(\lambda^+)$ implies $\mu \to^{\kappa(\lambda)} (\lambda^+)^{<\omega}_2$. So the consistency strength of these cardinals is in the realm of Erdős cardinals. Studying this hierarchy of cardinals further is somewhat out of our course so we continue now to look at Chang conjectures with more than four cardinals involved. In this case we need more Erdős-like cardinals. Similarly to Lemma 3.4 we can show the following.

LEMMA 3.9. (ZF) *Assume that $\lambda_0 < \lambda_1 < \cdots < \lambda_n$ and $\kappa_0 < \kappa_1 < \cdots < \kappa_n$ are cardinals such that $\kappa_i \to^{\kappa_{i-1}} (\lambda_i)^{<\omega}_2$. Then the Chang conjecture*

$$(\kappa_n, \ldots, \kappa_0) \twoheadrightarrow (\lambda_n, \ldots, \lambda_0) \text{ holds.}$$

PROOF. Let $\mathcal{A} = \langle \kappa_n, \ldots \rangle$ be an arbitrary first order structure in a countable language, and let

$$\{f_j\,;\, j \in \omega\}$$

be a complete set of Skolem functions for $\mathcal{A}$. Since $\kappa_n \to^{\kappa_{n-1}} (\lambda_n)_2^{<\omega}$ holds, let $I_n \in [\kappa_n \setminus \kappa_{n-1}]^{\lambda_n}$ be a set of good indiscernibles for $\mathcal{A}$. To take the next set of indiscernibles I_{n-1} we must make sure that it is, in a sense, compatible with I_n. That is, the Skolem hull of $I_n \cup I_{n-1}$ must not contain uncountably many elements below κ_{n-2}.

To do this we will enrich the structure $\mathcal{A}$ with functions, e.g.,

$$f_j(e_1, e_2, x_1, x_2),$$

for some f_j with arity $\mathsf{ar}(f_j) = 4$ and some $e_1, e_2 \in I_n$. Since f_j takes ordered tuples as arguments we must consider separately the cases $f_j(e_1, x_1, e_2, x_2)$, $f_j(e_1, x_1, x_2, e_2)$, etc..

Formally, let $\bar{I}_n \stackrel{\text{def}}{=} \{e_1, e_2, \ldots\}$ be the first ω-many elements of I_n. For every $s < \omega$ let $\{g_{s,t}\,;\, t < s!\}$ be an enumeration of all the permutations of s, and for every $t \in s!$ let

$$h_{s,t}(x_1, \ldots, x_s) \stackrel{\text{def}}{=} (x_{g_{s,t}(1)}, \ldots, x_{g_{s,t}}(s)).$$

For every $j < \omega$, every $k < \mathsf{ar}(f_j)$, and every $\ell \in \mathsf{ar}(f_j)!$ define a function $f_{j;k;\ell} : {}^{\mathsf{ar}(f_j)}\kappa_n \to \kappa_n$ by

$$f_{j;k;\ell}(x_1, \ldots, x_{\mathsf{ar}(f_j)-k}) \stackrel{\text{def}}{=} f_j(h_{\mathsf{ar}(f_j),\ell}(x_1, \ldots, x_{\mathsf{ar}(f_j)-k}, e_1, \ldots, e_k)).$$

Consider the structure

$$\mathcal{A}_{n-1} \stackrel{\text{def}}{=} \mathcal{A}^\frown \langle f_{i;c;t} \rangle_{j<\omega, k<\mathsf{ar}(f_j), \ell<\mathsf{ar}(f_j)!}.$$

Since $\kappa_{n-1} \to^{\kappa_{n-2}} (\lambda_{n-1})_2^{<\omega}$, let $I_{n-1} \in [\kappa_{n-1} \setminus \kappa_{n-2}]^{\lambda_{n-1}}$ be a set of good indiscernibles for $\mathcal{A}_{n-1}$.

Claim 1.
For any infinite set $Z \subseteq \kappa_{n-2}$ of size λ,

$$|\mathsf{Hull}_{\mathcal{A}}(I_n \cup I_{n-1} \cup Z) \cap \kappa_{n-2}| = \lambda.$$

PROOF OF CLAIM. Let $\bar{I}_{n-1} \stackrel{\text{def}}{=} \{e'_1, e'_2, \ldots\}$ be the first ω-many elements of I_{n-1}. The domain of $\mathsf{Hull}_{\mathcal{A}}(I_n \cup I_{n-1} \cup Z)$ is the set

$$X \stackrel{\text{def}}{=} \{f_j(\alpha_1, \ldots, \alpha_{\mathsf{ar}(f_j)})\,;\, j < \omega \text{ and } \alpha_1, \ldots, \alpha_{\mathsf{ar}(f_j)} \in I_n \cup I_{n-1} \cup Z\}.$$

If for some $x = f_j(\alpha_1, \ldots, \alpha_{\mathsf{ar}(f_j)}) \in X \cap \kappa_{n-2}$ there are elements of I_n among $\alpha_1, \ldots, \alpha_{\mathsf{ar}(f_j)}$ then since I_n is a set of indiscernibles for $\mathcal{A}$ and $\alpha_1, \ldots, \alpha_{\mathsf{ar}(f_j)}$ are finitely many, we can find $\alpha'_1, \ldots, \alpha'_{\mathsf{ar}(f_j)} \in \bar{I}_n \cup I_{n-1} \cup Z$ such that

$$x = f_j(\alpha_1, \ldots, \alpha_{\mathsf{ar}(f_j)}) = f_j(\alpha'_1, \ldots, \alpha'_{\mathsf{ar}(f_j)}).$$

We rewrite the tuple $(\alpha'_1, \ldots, \alpha'_{\mathsf{ar}(f_j)})$ so that the elements of $\bar{I}_n$ (if any) appear in ascending order at the end:

$$\{\alpha'_1, \ldots, \alpha'_n\} = \{\beta_1, \ldots, \beta_{\mathsf{ar}(f_j)-k}, e_1, \ldots, e_k\}.$$

Let $(\beta_1, \dots, \beta_{\mathsf{ar}(f_j)-k}, e_1, \dots, e_k)$ be a permutation of $(\alpha'_1, \dots, \alpha'_{\mathsf{ar}(f_j)})$, so for some $\ell < \mathsf{ar}(f_j)!$,

$$(\alpha'_1, \dots, \alpha'_{\mathsf{ar}(f_j)}) = h_{\mathsf{ar}(f_j),\ell}(\beta_1, \dots, \beta_{\mathsf{ar}(f_j)-k}, e_1, \dots, e_k).$$

But then

$$\begin{aligned} x &= f_j(\alpha'_1, \dots, \alpha'_{\mathsf{ar}(f_j)}) \\ &= f_j(h_{\mathsf{ar}(f_j),\ell}(\beta_1, \dots, \beta_{\mathsf{ar}(f_j)-k}, e_1, \dots, e_k) \\ &= f_{j,k,\ell}(\beta_1, \dots, \beta_{\mathsf{ar}(f_j)-k}). \end{aligned}$$

Therefore,

$$\begin{aligned} X \cap \kappa_{n-2} = \{f_{j,k,\ell}(\beta_1, \dots, \beta_{\mathsf{ar}(f_j)-k}) <\kappa_{n-2} \; ; \; j < \omega, k < \mathsf{ar}(f_j), \ell \in \mathsf{ar}(f_j)!, \\ \text{and } \beta_1, \dots, \beta_{\mathsf{ar}(f_j)-k} \in I_{n-1} \cup Z\}. \end{aligned}$$

But I_{n-1} is a set of good indiscernibles for $\mathcal{A}_{n-1}$, i.e., it is a set of indiscernibles for formulas with parameters below $\mathsf{min} I_{n-1} > \kappa_{n-2}$, therefore a set of indiscernibles for formulas with parameters from Z as well. Thus in the equation above we may replace I_{n-1} with $\bar{I}_{n-1}$. It's easy to see then that the set $X \cap \kappa_{n-2}$ has size λ. qed claim

Continuing like this we get for each $i = 1, \dots, n$ a set $I_i \in [\kappa_i \setminus \kappa_{i-1}]^{\lambda_i}$ of good indiscernibles for $\mathcal{A}$ with the property that for every infinite $Z \subseteq \kappa_{i-1}$, of size λ,

$$|\mathsf{Hull}_{\mathcal{A}}(I_n \cup \dots \cup I_i \cup Z) \cap \kappa_{i-1}| = \lambda.$$

So let $I \stackrel{\text{def}}{=} \bigcup_{i=1,\dots,n} I_i$ and take $\mathcal{B} \stackrel{\text{def}}{=} \mathsf{Hull}_{\mathcal{A}}(I \cup \lambda_0)$. By Lemma 0.23, we have that

$$\lambda_n = |I \cup \lambda_0| \leq |\mathsf{Hull}_{\mathcal{A}}(I \cup \lambda_0)| \leq |I \cup \rho| + \omega = \lambda_n.$$

Because for each $i = 1, \dots, n$ we have that $I_i \in [\kappa_i \setminus \kappa_{i-1}]^{\lambda_i}$ and by the way we defined the I_i, we have that

$$|\mathsf{Hull}(I \cup \lambda_0) \cap \kappa_i| = \lambda_i.$$

So the substructure $\mathsf{Hull}(I \cup \lambda_0) \prec \mathcal{A}$ is such as we wanted for our Chang conjecture to hold. qed

1.2. Infinitary Chang conjecture. Now let us consider the following infinitary version of the Chang conjecture, which turns out to be not as easy to handle.

DEFINITION 3.10. For cardinals $\kappa_0 < \dots < \kappa_n < \dots$ and $\lambda_0 < \dots < \lambda_n < \dots$, with $\kappa_n > \lambda_n$ for all n, define the infinitary Chang conjecture

$$(\kappa_n)_{n\in\omega} \twoheadrightarrow (\lambda_n)_{n\in\omega}$$

to mean that for every first order structure $\mathcal{A} = \langle \bigcup_{n\in\omega} \kappa_n, f_i, R_j, c_k \rangle_{i,j,k\in\omega}$ there is an elementary substructure $\mathcal{B} \prec \mathcal{A}$ with domain B of cardinality $\bigcup_{n\in\omega} \lambda_n$ such that for all $n \in \omega$, $|B \cap \kappa_n| = \lambda_n$.

Sometimes, when this uniform notation is not convenient, we will write the infinitary Chang conjecture as

$$(\dots, \kappa_n, \dots, \kappa_0) \twoheadrightarrow (\dots, \lambda_n, \dots, \lambda_0).$$

In [**For10**, §12 (4)] we read:

> Assuming that $2^{\aleph_0} < \aleph_\omega$, Silver showed that the cardinal $\aleph_\omega$ is Jónsson iff there is an infinite subsequence $\langle \kappa_n \,;\, n \in \omega\rangle$ of the $\aleph_n$'s such that the infinitary Chang conjecture of the form
>
> $$(\dots, \kappa_n, \kappa_{n-1}, \dots, \kappa_1) \twoheadrightarrow (\dots, \kappa_{n-1}, \kappa_{n-2}, \dots, \kappa_0)$$
>
> holds. It is not known how to get such a sequence of length 4.

All this is of course assuming AC. In the next section we will see how to get finite sequences of arbitrary finite length satisfying Chang conjectures in choiceless models of ZF.

For the infinitary version we need a coherent countable sequence of Erdős cardinals.

DEFINITION 3.11. Let $\lambda_1 < \dots < \lambda_i < \dots$ and $\kappa_0 < \dots < \kappa_i < \dots$ be cardinals and let $\kappa \stackrel{\text{def}}{=} \bigcup_{i<\omega} \kappa_i$. We say that the sequence $\langle \kappa_i \,;\, i < \omega\rangle$ is a coherent sequence of Erdős cardinals with respect to $\langle \lambda_i \,;\, 0 < i < \omega\rangle$ if for every $\gamma < \kappa_1$ and every $f : [\kappa]^{<\omega} \to \gamma$ there is a sequence $\langle A_i \,;\, 0 < i < \omega\rangle$ such that

(1) for every $0 < i < \omega$, $A_i \in [\kappa_i \setminus \kappa_{i-1}]^{\lambda_i}$, and
(2) if $x, y \in [\kappa]^{<\omega}$ are such that $x, y \subseteq \bigcup_{i<\omega} A_i$ and for every $0 < i < \omega$ $|x \cap A_i| = |y \cap A_i|$ then $f(x) = f(y)$.

Such a sequence $\langle A_i \,;\, 0 < i < \omega\rangle$ is called a $\langle \lambda_i \,;\, 0 < i < \omega\rangle$-coherent sequence of homogeneous sets for f with respect to $\langle \kappa_i \,;\, i < \omega\rangle$. Note that the 0th element of a coherent sequence of Erdős cardinals need not be an Erdős cardinal.

Coherent sequences of Ramsey cardinals are as such sequences. In [**AK06**, Theorem 3] a coherent sequence of Ramsey cardinals is forced from a model of ZFC with one measurable cardinal. Similarly to [**AK08**, Proposition 8] we get the following lemma which says that coherent sequences of Erdős cardinals give coherent sequences of indiscernibles.

LEMMA 3.12. *Let $\lambda_1 < \dots < \lambda_i < \dots$ and $\kappa_0 < \dots < \kappa_i < \dots$ be cardinals and let $\kappa \stackrel{\text{def}}{=} \bigcup_{i<\omega} \kappa_i$. If the sequence $\langle \kappa_i \,;\, i < \omega\rangle$ is a coherent sequence of Erdős cardinals with respect to $\langle \lambda_i \,;\, 0 < i < \omega\rangle$ then for every first order structure $\mathcal{A} = \langle \kappa, \dots\rangle$ with a countable language, there is a $\langle \lambda_i \,;\, 0 < i < \omega\rangle$-coherent sequence of good indiscernibles for $\mathcal{A}$ with respect to $\langle \kappa_i \,;\, i < \omega\rangle$, i.e., a sequence $\langle A_i \,;\, 0 < i < \omega\rangle$ such that*

(1) for every $0 < i < \omega$, $A_i \in [\kappa_i \setminus \kappa_{i-1}]^{\lambda_i}$, and
(2) if $x, y \in [\kappa]^{<\omega}$ are such that $x = \{x_1, \dots, x_n\}$, $y = \{y_1, \dots, y_n\}$, $x, y \subseteq \bigcup_{0<i<\omega} A_i$, and for every $0 < i < \omega$ $|x \cap A_i| = |y \cap A_i|$ then for every $(n+\ell)$-ary formula φ in the language of $\mathcal{A}$ and every $z_1, \dots, z_\ell < \min\{x_1, \dots, x_n, y_1, \dots, y_n\}$,

$$\mathcal{A} \models \varphi(z_1, \dots, z_\ell, x_1, \dots, x_n) \iff \mathcal{A} \models \varphi(z_1, \dots, z_\ell, y_1, \dots, y_n).$$

With this and the proof of Lemma 3.9 in mind we can prove the following.

LEMMA 3.13. (ZF) *If $\lambda_1 < \dots < \lambda_n < \dots$ and $\kappa_0 < \dots < \kappa_i < \dots$ are cardinals such that $\langle \kappa_i \,;\, i < \omega\rangle$ is a coherent sequence of Erdős cardinals with respect to $\langle \lambda_i \,;\, 0 < i < \omega\rangle$, then for* *any* $\lambda_0 < \lambda_1$*, the Chang conjecture*

$$(\kappa_n)_{n<\omega} \twoheadrightarrow (\lambda_n)_{n\in\omega}$$

holds.

Again, let us connect this to the easier Erdős-like cardinals $\kappa \to^{\theta} (\lambda)_2^{<\omega}$. For this let's define first what it means to be a coherent sequence of cardinals with that property.

DEFINITION 3.14. Let $\langle \kappa_i \,;\, i < \omega \rangle$ and $\langle \lambda_i \,;\, 0 < i < \omega \rangle$ be increasing sequences of cardinals. Let $\kappa = \bigcup_{i<\omega} \kappa_i$. We say that the sequence $\langle \kappa_i \,;\, i < \omega \rangle$ is a coherent sequence of cardinals with the property $\kappa_{i+1} \to^{\kappa_i} (\lambda_{i+1})_2^{<\omega}$ iff for every structure $\mathcal{A} = \langle \kappa, \ldots \rangle$ with a countable language, there is a $\langle \lambda_i \,;\, 0 < i < \omega \rangle$-coherent sequence of good indiscernibles for $\mathcal{A}$ with respect to $\langle \kappa_i \,;\, i < \omega \rangle$.

As before we can prove the following.

COROLLARY 3.15. (ZF) *Let* $\langle \kappa_n \,;\, n < \omega \rangle$ *and* $\langle \lambda_n \,;\, 0 < n < \omega \rangle$ *be increasing sequences of cardinals, and let* $\kappa = \bigcup_{n<\omega} \kappa_n$*. If* $\langle \kappa_i \,;\, i < \omega \rangle$ *is a coherent sequence of cardinals with the property* $\kappa_{n+1} \to^{\kappa_n} (\lambda_{n+1})_2^{<\omega}$ *then the Chang conjecture*

$$(\kappa_n)_{n\in\omega} \twoheadrightarrow (\lambda_n)_{n\in\omega}$$

holds.

.

1.3. Weak Chang conjecture and almost $<\tau$-Erdős cardinals. A weaker version of the Chang conjectures was considered in [**She80**, §35] in its combinatorial form, and studied extensively in [**DK83**] in its model theoretic form. This is a generalised version of it.

DEFINITION 3.16. For a successor cardinal τ, and a cardinal $\kappa > \tau$, the weak Chang conjecture $\mathsf{wCc}(\kappa, \tau)$ means that for every structure $\mathcal{A} = \langle \kappa, \ldots \rangle$ with a countable language and every $\xi < \tau$, there is an $\alpha \in (\xi, \tau)$ and a sequence $\langle \mathcal{B}_\beta, \sigma_{\gamma\beta} \,;\, \gamma \leq \beta < \tau \rangle$ such that for all $\gamma \leq \beta < \tau$,

(i) $\mathcal{B}_\beta \prec \mathcal{A}$,
(ii) $\xi \subseteq \mathcal{B}_\beta \cap \tau = \mathcal{B}_\gamma \cap \tau \subseteq \alpha$,
(iii) $\mathsf{ot}(\mathcal{B}_\beta) > \beta$, and
(iv) the map $\sigma_{\gamma\beta} : \mathcal{B}_\gamma \to \mathcal{B}_\beta$ is elementary, with $\sigma_{\gamma\beta}(\mu) = \mu$ for every $\mu \in \mathcal{B}_\beta \cap \tau$.

Such a system of substructures may look as follows (in the image we assume that $\mathcal{A} = \langle L_\kappa, \ldots \rangle$ for a better visual).

The weak Chang conjecture $\mathsf{wCc}(\kappa, \tau)$ follows trivially from the Chang conjecture $(\kappa, \tau) \twoheadrightarrow (\tau, \alpha)$. In ZFC the we have the following equivalent versions of the weak Chang conjecture.

LEMMA 3.17. (ZFC) *If* $\kappa > \tau$ *and* τ *is a successor cardinal, then the following are equivalent.*

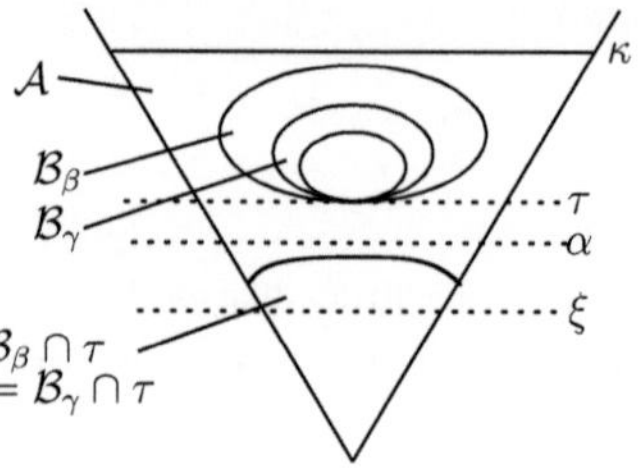

(1) $\mathsf{wCc}(\tau^+, \tau)$

(2) *For every first order structure* $\mathcal{A} = \langle A, \ldots \rangle$ *with a countable language, if* $\tau^+ \subseteq A$ *then there is* $\alpha < \tau$ *such that for all* $\beta < \tau$ *there is* $X \prec \mathcal{A}$ *with* $X \cap \tau \subseteq \alpha$ *and* $\mathsf{ot}(X \cap \tau^+) > \beta$.

(3) *For every first order structure* $\mathcal{A} = \langle A, \ldots \rangle$ *with a countable language, if* $\tau^+ \subseteq A \subseteq H_{\tau^+}$*, then for every* $\xi < \tau$ *there exists an elementary map* $\pi : \bar{A}$ *such that* $\bar{A}$ *is transitive, the critical point* $\mathsf{crit}(\pi) = \alpha \in (\xi, \tau)$*, and for every* $\beta < \tau$ *there exists an elementary map* $\pi' : \mathcal{B}' \to \mathcal{A}$ *such that* $\mathcal{B}'$ *is transitive,* $\bar{A} \prec \mathcal{B}'$*,* $\pi' \restriction \bar{A} = \pi$*, and* $\mathrm{Ord} \cap \mathcal{B}' > \beta$.

(4) *Statement* (3) *and moreover if* $\langle \mathcal{B}'_\beta \, ; \, \beta < \tau \rangle$ *is the sequence we get from the last requirement of* (3)*, then for every* $\gamma \leq \beta < \tau$*,* $\mathcal{B}'_\gamma \prec \mathcal{B}'_\beta$.

PROOF. The implications (1) $\Rightarrow$ (2), (4) $\Rightarrow$ (1), and (4) $\Rightarrow$ (3) are trivial. The equivalence (2) $\Leftrightarrow$ (3) is [**DK83**, Theorem 5.1]. To show (2) $\Rightarrow$ (4), just follow the proof of [**DK83**, Theorem 5.1] and note that when "$\mathsf{wCC}^*(\tau)$" (i.e., (3)) is derived, the transitive substructures are a chain of substructures as well (second to last line of that proof). qed

One could ask whether this result could be proven in ZF alone, with A replaced by κ in both (2) and (3). A careful reading of the proof of [**DK83**, Theorem 5.1] shows that for the step (2) $\Rightarrow$ (3), AC_ω and AC_{τ^+} are used. This of course doesn't mean that it is impossible to find another proof in ZF. This equivalence in ZF remains an open question.

As with the Chang conjectures, there's a weakening of the notion of an Erdős cardinal that is connected with the weak Chang conjecture.

DEFINITION 3.18. For a function $f : [S]^{<\omega} \to V$, $S \subseteq \mathrm{Ord}$, and an infinite set $X \subseteq S$ that is homogeneous for f, we define

$$\mathsf{tp}_f(X) \stackrel{\text{def}}{=} \langle y_n \, ; \, \exists \gamma_1 < \cdots < \gamma_n \in X(f(\gamma_1, \ldots, \gamma_n) = y_n) \wedge n \in \omega \rangle.$$

This set is called the type of X with respect to f. It's the sequence of all the n-colours of X via f.

A sequence $\langle X_\beta \, ; \, \beta < \tau \rangle$ is called a homogeneous sequence for f of order τ if all the sets in the sequence are homogeneous with respect to f, all have the same type, and for every $\beta < \tau$, $\mathsf{ot}(X_\beta) = \omega(1 + \beta)$.

If $\tau = \omega\tau \neq 0$ and $\lambda < \kappa$ are all ordinals then define

$$\kappa \to (<\tau)^{<\omega}_{\lambda}$$

iff for every $f : [\kappa]^{<\omega} \to \lambda$ there is a homogeneous sequence $\langle X_\alpha \,;\, \alpha < \tau\rangle$ for f.

A cardinal κ is the almost $<\tau$-Erdős iff it is regular and it's the least such that $\kappa \to (<\tau)^{<\omega}_2$ holds.

This definition is slightly different than the definition of the almost $<\tau$-Erdős that is used in [**DK83**, page 235] to derive the weak Chang conjecture, but these two are easily seen to be equiconsistent. As noted in [**DK83**, §8, Fact 1], ZFC implies that the almost $<\tau$-Erdős cardinal is inaccessible but as mentioned right before that fact, it is not necessarily Mahlo. As with the Erdős cardinals we have the following.

PROPOSITION 3.19. (ZFC) *If κ is almost $<\tau$-Erdős then for every $\gamma < \tau$*

$$\kappa \to (<\tau)^{<\omega}_{\gamma}$$

holds.

Therefore by [**DK83**, §8, Fact 1], our notion of the almost $<\tau$-Erdős cardinal also implies inaccessibility. We will need this fact later to construct the Jech model for the almost $<\tau$-Erdős cardinal. As with Erdős cardinals, almost $<\tau$-Erdős cardinals imply the existence of sets of good indiscernibles.

PROPOSITION 3.20. (ZF) *The partition property $\kappa \to (<\tau)^{<\omega}_2$ is equivalent to the following: for every first order structure $\mathcal{A} = \langle A, \ldots\rangle$ with a countable language and such that $\kappa \subseteq A$ there is a sequence $\langle X_\beta \,;\, \beta < \tau\rangle$ such that for every $\gamma, \beta < \tau$, X_β is a set of indiscernibles for $\mathcal{A}$ such that $\mathsf{ot}(X_\beta) = \omega(1+\beta)$ and X_γ, X_β agree in the formulas of $\mathcal{A}$, i.e., for every n-ary formula φ in the language of $\mathcal{A}$, every $x_1 < \cdots < x_n \in X_\beta$ and every $y_1 < \cdots < y_n \in X_\gamma$,*

$$\mathcal{A} \models \varphi(x_1, \ldots, x_n) \iff \mathcal{A} \models \varphi(y_1, \ldots, y_n).$$

The proof of this is exactly as Silver's proof of [**Kan03**, Theorem 9.3], with the added part that the sets of indiscernibles agree on the formulas of $\mathcal{A}$ which is straightforward considering the sets X_β have the same types.

Also with Proposition 3.20 and the same proof as [**AK08**, Proposition 8] we get the following.

PROPOSITION 3.21. (ZF) *If κ is almost $<\tau$-Erdős then for any first order structure $\mathcal{A} = \langle A, \ldots\rangle$ with a countable language, there is a sequence $\langle X_\beta \,;\, \beta < \tau\rangle$ such that for every $\beta, \gamma < \tau$, X_β is a set of good indiscernibles for $\mathcal{A}$, $\mathsf{ot}(X_\beta) > \beta$, and X_β, X_γ agree in the formulas of $\mathcal{A}$.*

An almost $<\tau$-Erdős cardinal implies the weak Chang conjecture.

LEMMA 3.22. (ZF) *If τ is a regular cardinal, and $\kappa > \tau$ is the almost $<\tau$-Erdős cardinal, then $\mathsf{wCc}(\kappa, \tau)$ holds.*

PROOF. Let $\mathcal{A} = \langle \kappa, \ldots \rangle$ be an arbitrary structure. Since κ is minimal such that $\kappa \to (<\tau)_2^{<\omega}$, let $g : [\tau]^{<\omega} \to 2$ be such that it has no homogeneous sequences of order τ. Consider the structure

$$\bar{\mathcal{A}} = \mathcal{A}^\frown \langle \tau, g\restriction[\tau]^n \rangle_{n<\omega},$$

where τ, $g\restriction[\tau]^n$ are viewed as relations. By Proposition 3.21 there is a sequence $\langle X_\beta \,;\, \beta < \tau \rangle$ such that for every $\beta, \gamma < \tau$, X_β is a set of good indiscernibles for $\bar{\mathcal{A}}$, $\mathsf{ot}(X_\beta) > \beta$, and X_γ, X_β agree on the formulas of $\bar{\mathcal{A}}$. Since g has no homogeneous sequences of order τ there must be a $\beta < \tau$ such that $X_\beta \not\subseteq \tau$. By indiscernibility, $X_\beta \subseteq \kappa \setminus \tau$. Since for every $\gamma, \beta < \tau$, X_β and X_γ agree in the formulas of $\mathcal{A}$, for every $\beta < \tau$, $X_\beta \subseteq \kappa \setminus \tau$.

For every $n < \omega$ let $\bar{X}_n$ be the first ω-many elements of X_n and for every $\omega \leq \zeta < \tau$ let $\bar{X}_\zeta$ be the first ζ-many elements of X_ζ.

Let $\xi < \tau$ be arbitrary. For every $\beta < \tau$ let

$$\mathcal{B}_\beta \overset{\text{def}}{=} \mathsf{Hull}_{\mathcal{A}}(X_\beta \cup \xi),$$

and for $\gamma \leq \beta < \tau$ let $\sigma_{\gamma\beta} : \bar{X}_\gamma \to \bar{X}_\beta$ be the function that sends the ζ'th element of $\bar{X}_\gamma$ to the ζ'th element of $\bar{X}_\beta$. Similarly to the Elementary Embedding Theorem [**CK90**, Theorem 3.3.11(d)] we get that $\sigma_{\gamma\beta}$ can be extended uniquely to an elementary embedding $\bar{\sigma}_{\gamma\beta} : \mathcal{B}_\gamma \to \mathcal{B}_\beta$ by sending a $y \in \mathcal{B}_\beta$ such that

$$y = t(\xi_1, \ldots, \xi_n, x_1, \ldots, x_r)$$

(where t is a term of the language of $\mathcal{A}$, $\xi_1, \ldots, \xi_n \in \xi$, and $x_1, \ldots, x_r \in X_\gamma$) to

$$\bar{\sigma}_{\gamma\beta}(y) \overset{\text{def}}{=} t(\xi_1, \ldots, \xi_n, \sigma_{\gamma\beta}(x_1), \ldots, \sigma_{\gamma\beta}(x_n)).$$

Note that we don't use the axiom of choice for this because the language of $\mathcal{A}$ and all the sets involved are wellorderable.

Now for every $\beta < \tau$, because $\bar{X}_\beta$ is a set of good indiscernibles, the elements of $\bar{X}_\beta$ are indiscernibles with parameters $< \tau$. So

$$|\mathcal{B}_\beta \cap \tau| \leq \xi\omega = \xi.$$

So there is some $\alpha \in (\xi, \tau)$ such that $\xi \subseteq \mathcal{B}_\beta \cap \tau \subseteq \alpha$. Fix α. Now it suffices to show that for every $\beta, \gamma < \tau$, $\mathcal{B}_\beta \cap \tau = \mathcal{B}_\gamma \cap \tau$. So let $\beta, \gamma < \tau$ and $x \in \mathcal{B}_\gamma \cap \tau$. Without loss of generality assume that for some function f_i and some $\xi_1, \ldots, \xi_\ell < \tau$ and some $\zeta_1 < \cdots < \zeta_r \in X_\gamma$,

$$\mathcal{A} \models x = f(\xi_1, \ldots, \xi_\ell, \zeta_1, \ldots, \zeta_r).$$

But this is equivalent to

$$\mathcal{A} \models x = f(\xi_1, \ldots, \xi_\ell, \zeta_1', \ldots, \zeta_r'),$$

for any $\zeta_1' < \cdots < \zeta_r' \in X_\beta$, i.e., $x \in \mathcal{B}_\beta \cap \tau$. Note that since we have this, by the construction of $\sigma_{\beta\gamma}$ we also get that for every $\mu \in \mathcal{B}_\beta \cap \tau$, $\sigma_{\gamma\beta}(\mu) = \mu$. qed

As with the Chang conjectures, we notice that to prove this we only need the following property.

DEFINITION 3.23. For τ regular cardinal and $\kappa > \tau$ define

$$\kappa \rightarrow^\tau (<\tau)^{<\omega}_2$$

to mean that for every structure $\mathcal{A} = \langle \kappa, \ldots \rangle$ with a countable language, there is a sequence $\langle I_\beta \,;\, \beta < \tau \rangle$ such that for every $\beta < \tau$ the set $I_\beta \subseteq \kappa \setminus \tau$ is a set of indiscernibles with respect to parameters $< \tau$, $\mathsf{ot}(I_\beta) > \beta$, and for every $\beta, \gamma < \tau$, I_β and I_γ agree on the formulas of $\mathcal{A}$.

Immediately we get the following.

COROLLARY 3.24. (ZF) *If τ is a regular cardinal, and $\kappa > \tau$ then $\kappa \rightarrow^\tau (<\tau)^{<\omega}_2$ implies* $\mathsf{wCc}(\kappa, \tau)$.

2. The Dodd-Jensen core model and HOD

In this section we will define the Dodd-Jensen core model K^{DJ} and give a list of its properties that are commonly used as black boxes. We will use these properties in the next section to get indiscernibles from our Chang conjectures.

We use the exposition of the Dodd-Jensen core model given in [**DK83**] which is simple and sufficient for what we need for the next section. We repeat the definitions, and the lemmas we will use as black boxes. We assume some familiarity with constructibility theory. We will start with the J-hierarchy.

DEFINITION 3.25. If U is a set, a function f is called rudimentary in U iff it is finitely generated by the following operations.

(1) $f(x_1, \ldots, x_n) = x_1$,
(2) $f(x_1, \ldots, x_n) = x_i \setminus x_j$,
(3) $f(x_1, \ldots, x_n) = \{x_i, x_j\}$,
(4) $f(x_1, \ldots, x_n) = h(g(x_1, \ldots, x_n))$,
(5) $f(y, x_1, \ldots, x_n) = \bigcup_{z \in y} g(z, x_1, \ldots, x_n)$, and
(6) $f(x_1, \ldots, x_n) = x_i \cap U$.

The closure of $x \cup \{x\}$ under functions that are rudimentary in U is denoted by $\mathsf{rud}_U(x)$. The J-hierarchy is defined as follows.

$$J^U_0 \stackrel{\text{def}}{=} \varnothing,$$
$$J^U_{\alpha+1} \stackrel{\text{def}}{=} \mathsf{rud}_U(J^U_\alpha), \text{ and}$$
$$J^U_\lambda \stackrel{\text{def}}{=} \bigcup_{\alpha<\lambda} J^U_\alpha, \text{ for a limit ordinal } \lambda.$$

The basic building blocks for the core model are iterable premice.

DEFINITION 3.26. Let $\alpha \leq \mathrm{Ord}$. A structure $M = J^U_\alpha$ is a premouse at μ iff

$$M \models U \text{ is a normal measure over } \mu > \omega.$$

This M is called ξ-iterable ($\xi \leq \mathrm{Ord}$) iff there is a system $\langle M_i, \pi_{ij}, \mu_i, U_i \rangle_{i \leq j < \xi}$

(1) $M_0 = M$,

(2) for every $i < \xi$, M_i is a premouse at μ_i with measure U_i,
(3) for every $i \leq j < \xi$, $\pi_{ij} : M_i \to M_j$ is a Σ_1-embedding, and $\pi_{ii} = \mathsf{id}\restriction M_i$,
(4) the system $\langle \pi_{ij} \rangle_{i \leq j < \xi}$ commutes,
(5) if $i+1 < \xi$ then $M_{i+1} = \mathsf{Ult}_{U_i}(M_i)$ and $\pi_{i,i+1}$ is the embedding coming from the ultrapower construction, and
(6) if $\zeta < \xi$ is a limit ordinal then $\langle M_\zeta, \pi_{i\zeta} \rangle$ is the transitive direct limit of $\langle M_i, \pi_{ij} \rangle_{i \leq j < \zeta}$.

If such a system exists for M then it is called the ξ-iteration of M. If M is Ord-iterable then we call M iterable.

The property of being an iterable premouse is described by a formula therefore it is transferred via elementary embeddings. In certain cases we can get these images of iterable premice to be iterable premice in the universe. For this we use the following, which is [**DK83**, Lemma 1.16].

LEMMA 3.27. (ZFC) *Let A be a transitive model of* ZF^{--}, *i.e.,* ZF *without powerset and replacement. If $\omega_1 \subseteq A$ then an iterable premouse in A is an iterable premouse in the universe.*

The next lemma is used to transfer the property of being an iterable premouse via Σ_1-elementary embeddings. It can be found in [**DK83**, Lemma 1.17].

LEMMA 3.28. (ZFC) *Let $\sigma : \bar{M} \to M$ be a Σ_1-embedding, where M is an η-iterable premouse and $\bar{M}$ is transitive. Then $\bar{M}$ is a premouse and it's η-iterable.*

Now to define the Dodd-Jensen core model K^{DJ}.

DEFINITION 3.29. For an iterable premouse M at μ, define the lower part of M as $\mathsf{lp}(M) \overset{\text{def}}{=} M \cap V_\mu$. The Dodd-Jensen core model K^{DJ} is the class

$$K^{\mathsf{DJ}} \overset{\text{def}}{=} L \cup \bigcup \{\mathsf{lp}(M) \ ; \ M \text{ is an iterable premouse}\}.$$

According to [**DK83**, end of page 241], in ZF^{--} (i.e., ZF without powerset and without replacement) this definition yields the same core model as the one defined by Dodd and Jensen. This is [**DJ81**, Definition 6.3] which gives a class D such that $K^{\mathsf{DJ}} = L[D]$. For any cardinal α of K^{DJ} define

$$K^{\mathsf{DJ}}_\alpha \overset{\text{def}}{=} H^{K^{\mathsf{DJ}}}_\alpha,$$

i.e., the sets of hereditary cardinality below α in K^{DJ}.

In [**DJ81**, Lemma 6.9] we find the following.

LEMMA 3.30. (ZFC) *If $\beta \geq \omega$ is a cardinal in K^{DJ} then*

$$J^D_\beta = H^{K^{\mathsf{DJ}}}_\beta = K^{\mathsf{DJ}}_\beta.$$

In [**DK83**, Lemma 2.1] we find the following useful lemma.

LEMMA 3.31. (ZFC) *If α is an uncountable cardinal in K^{DJ} then K^{DJ}_α models "$V = K^{\mathsf{DJ}}$" in the sense of the above definition of K^{DJ} and $K^{\mathsf{DJ}}_\alpha \models \mathsf{ZF}^{--}$, i.e.,* ZF *without powerset and without replacement.*

The following is [**DK83**, Lemma 2.8] and it is used to prove that certain ultrapowers are well-founded and thus get the existence of elementary embeddings from K^{DJ} to K^{DJ}.

LEMMA 3.32. (ZFC) *Let U be an ultrafilter on $\mathcal{P}(\alpha) \cap K^{\mathsf{DJ}}$, and let λ be a cardinal such that $\lambda > \alpha$ and $\lambda \geq \omega_1$. Assume that the ultrapower $({}^{\alpha}K^{\mathsf{DJ}} \cap K^{\mathsf{DJ}})/U$ is not well-founded. Then there are $f_0, f_1, \cdots \in K^{\mathsf{DJ}}_{\lambda}$ such that for every $i < \omega$,*

$$\{\nu < \alpha \,;\, f_{i+1}(\nu) \in f_i(\nu)\} \in U.$$

Lemma 2.9 of [**DK83**] says that if $\lambda \geq \omega_1$ then an elementary embedding of $K^{\mathsf{DJ}}_{\lambda}$ to itself extends to an elementary embedding of K^{DJ} to itself. In fact the proof yields the following stronger version.

LEMMA 3.33. (ZFC) *If $\lambda \geq \omega_1$ and $\pi : K^{\mathsf{DJ}}_{\lambda} \to M$ is an elementary embedding with critical point α and M is well founded and transitive, then there is an elementary embedding $\tilde{\pi} : K^{\mathsf{DJ}} \to K^{\mathsf{DJ}}$ with critical point α.*

In [**DK83**, 1.4] we see that an elementary embedding from K^{DJ} to a transitive class is an elementary embedding from K^{DJ} to K^{DJ}.

LEMMA 3.34. (ZFC) *Let $\pi : K^{\mathsf{DJ}} \to M$ be elementary and M be transitive. Then $M = K^{\mathsf{DJ}}$.*

In [**DK83**, 1.5] we find the following useful way to connect elementary embeddings of K^{DJ} to K^{DJ} with inner models of measurable cardinals.

LEMMA 3.35. (ZFC) *Let $\pi : K^{\mathsf{DJ}} \to K^{\mathsf{DJ}}$ be nontrivial and elementary. Let α be the first ordinal moved by π. Then there is an inner model with a measurable cardinal β, such that*

- *if $\alpha < \omega_1$ then $\beta \leq \omega_1$, and*
- *if $\alpha \geq \omega_1$ then $\beta < \alpha^+$.*

These inner models with measurable cardinals are of the form $L[U]$, where U is such that $L[U] \models$"U is a normal ultrafilter on κ". The following is [**Kan03**, Exercise 20.1].

PROPOSITION 3.36. (ZFC) *Let κ be a measurable cardinal and U a κ-complete ultrafilter over κ. Let $\bar{U} \stackrel{\text{def}}{=} U \cap L[U] \in L[U]$. Then*

$$L[U] \models \bar{U} \textit{ is a } \kappa\textit{-complete ultrafilter over } \kappa$$

and if U is normal then $L[U] \models \bar{U}$ is normal.

In [**DK83**, 1.6] we find the following way to connect these structures with K^{DJ}.

LEMMA 3.37. (ZFC) *Assume $L[U] \models$"U is a normal ultrafilter on κ". Then $\mathcal{P}(\kappa) \cap K^{\mathsf{DJ}} = \mathcal{P}(\kappa) \cap L[U]$. This implies that $V_\kappa \cap K^{\mathsf{DJ}} = V_\kappa \cap L[U]$, and further that $K^{\mathsf{DJ}} = \bigcap_{i<\infty}(L[U])_i$, where $(L[U])_i$ is the i-th iterated ultrapower of $L[U]$.*

So we need sets $L_\varsigma[U]$ that are iterable premice. For this lemma above to be useful we note the following lemma which is [**Kan03**, Lemma 20.5].

LEMMA 3.38. (ZFC) *Suppose that* $\langle L_\varsigma[U], \in, U\rangle$ *is a premouse,* $L_\varsigma[U] \models \mathsf{ZFC}$, *and* $\omega_1 \cup \{U\} \subseteq L_\varsigma[U]$. *Then* $\langle L_\varsigma[U], \in, U\rangle$ *is iterable.*

A structure $\langle L[U], \in, U\rangle$ that is a premouse such that $U \in L[U]$ and $L[U] \models$"U is a normal measure over κ", is called a κ-model. The lemma below is [**Kan03**, Corollary 20.7], and it is used to show that if an ordinal κ is measurable in some inner model, then every regular cardinal greater than κ^+ is measurable in an inner model.

LEMMA 3.39. (ZFC) *Suppose that there is a* κ*-model, and* ν *is a regular cardinal greater than* κ^+. *Set* $\bar{C}_\nu = C_\nu \cap L[C_\nu]$, *where* C_ν *is the closed unbounded filter over* ν. *Then* $L[\bar{C}_\nu]$ *is a* ν*-model, i.e.,*

$$\langle L[\bar{C}_\nu], \in, \bar{C}_\nu\rangle \models \bar{C}_\nu \textit{ is a normal ultrafilter over } \nu.$$

Back to general premice, their iterations can be linearly ordered as we see in the next lemma, which is [**DK83**, Lemma 1.13].

LEMMA 3.40. (ZFC) *Let* M, N *be iterable premice and* θ *a regular cardinal above* $|M|$ *and* $|N|$. *Then either* $M_\theta \in N_\theta$, $M_\theta = N_\theta$, *or* $N_\theta \in M_\theta$.

To get indiscernibles from iterating premice we will use the following lemma, which we can find in [**DK83**, Lemma 1.14].

LEMMA 3.41. (ZFC) *Let* $\mathcal{M} = \langle M_i, \pi_{ij}, \kappa_i, U_i\rangle_{i \leq j < \eta}$ *be the* η*-iteration of an* η*-iterable premouse* M. *Then for* $j < \eta$:

(1) $M_j = \{\pi_{oj}(f)(\kappa_{i_1}, \ldots, \kappa_{i_n})\ ;\ n < \omega,\ f : \kappa_0^n \to M_0,$ *and* $i_1 < \cdots < i_n < j\}$.
(2) *For* φ *a* Σ_0 *formula in the language of* M_0, $x \in M_0$, *and* $i_1 < \ldots, i_n < j$, *we have that* $M_j \models \varphi[\pi_{0j}(x), \kappa_{i_1}, \ldots, \kappa_{i_n}]$ *iff*

$$\exists X \in U_0 \cap M_0 \forall x_1, \ldots, x_n \in X (x_1 < \cdots < x_n \to M_0 \models \varphi[x, x_1, \ldots, x_n].$$

(3) $\{x_i\ ;\ i < j\}$ *is a set of indiscernibles for* $\langle M_j, \langle \pi_{oj}(x)\ ;\ x \in M_0\rangle\rangle$.

To get sets of indiscernibles in K^{DJ} we need Jensen's indiscernibility lemma as stated in [**DJK79**, Lemma 1.3].

LEMMA 3.42. (ZFC) *Let* $A \subseteq \kappa$ *be such that* $L_\kappa[A] \subseteq K_\kappa^{\mathsf{DJ}}$, *and consider the structure* $\mathcal{A} = \langle K_\kappa^{\mathsf{DJ}}, \in, D \cap \kappa, A\rangle$. *Let* I *be a set of good indiscernibles for* $\mathcal{A}$ *such that* $\mathsf{cf}(\mathsf{ot}(I)) > \omega$. *Then there is some* $I' \in K^{\mathsf{DJ}}$ *such that* $I' \in K^{\mathsf{DJ}}$, $I \subseteq I'$, *and* I' *is good set of indiscernibles for* $\mathcal{A}$.

Note that the K_κ^{DJ} are defined differently in [**DJK79**] but by Lemma 3.30 this definition is equivalent to the one we use here. As all the results in this section, this is an equivalence in ZFC. So, in order for K^{DJ} to have all these nice properties, we will build K in the universe of hereditarily ordinal definable sets.

DEFINITION 3.43. A set X is called ordinal definable iff there is a formula φ with $n+1$ many free variables, and there are ordinals $\alpha_1, \ldots, \alpha_n$ such that for every x,

$$x \in X \text{ iff } \varphi(\alpha_1, \ldots, \alpha_n, x).$$

This is not a formal definition but it can be made into one (see [**Jec03**, 13.25, 13.26, and Lemma 13.25]). The class of all hereditarily ordinal definable sets is defined as

$$\mathsf{HOD} \stackrel{\text{def}}{=} \{X \ ; \mathsf{tc}(\{X\}) \subset \mathsf{OD}\}.$$

(See also [**Kun80**, Chapter V, §2], [**Jec03**, page 194]). By [**Jec03**, Theorem 13.26] HOD is a transitive model of ZFC.

For sets A and X we say that X is ordinal definable from A iff there is a formula φ with $n+2$ many free variables, and there are ordinals $\alpha_1, \ldots, \alpha_n$ such that

$$x \in X \text{ iff } \varphi(A, \alpha_1, \ldots, \alpha_n, x).$$

Again, there is a formal definition for this notion (see [**Jec03**, 13.27 and 13.28]). The class of all sets that are hereditarily ordinal definable from A is denoted by $\mathsf{HOD}[A]$ and according to the comments following [**Jec03**, 13.28], $\mathsf{HOD}[A]$ is also a transitive model of ZFC.

This is a very useful model we can build in any model of ZF. Because the results concerning the Dodd-Jensen core model require K^{DJ} to be built in a model of ZFC we will use some $\mathsf{HOD}[A]$ to build K^{DJ} and use the following folklore result.

LEMMA 3.44. (ZF) *For any set of ordinals x, $(K^{\mathsf{DJ}})^{\mathsf{HOD}} = (K^{\mathsf{DJ}})^{\mathsf{HOD}[x]}$ and this equality is in every level of the K^{DJ} construction.*

For a proof see [**AK06**, Proposition 1.1].

3. Successor of a regular

In Section 3 of Chapter 1 we saw how to construct models of $\mathsf{ZF} + \neg\mathsf{AC}$ in which a "small" large cardinal becomes the successor of some regular cardinal, while retaining its large cardinal properties. This is the model we will be using for the forcing side. For the core model side we will modify existing ZFC arguments into $\mathsf{ZF} + \neg\mathsf{AC}$ arguments, by using Lemma 3.44.

3.1. Forcing side. We will start with the four cardinal Chang conjecture. We begin from an Erdős cardinal $\kappa(\lambda)$ in ZFC, we collapse it symmetrically to become η^+ for some regular cardinal η, and we end up with a model of ZF+"for every $\theta < \kappa$ and every $\rho < \lambda$, $(\eta^+, \theta) \twoheadrightarrow (\lambda, \rho)$. To do this we will use the Erdős-like property $\kappa \to^{\theta} (\lambda)_2^{<\omega}$.

LEMMA 3.45. *If V is a model of ZFC+ "$\kappa = \kappa(\lambda)$ exists", $\mathbb{P}$ is a partial order such that $|\mathbb{P}| < \kappa$, and G is a $\mathbb{P}$ generic filter, then in $V[G]$ for any $\theta < \kappa$, $\kappa \to^{\theta} (\lambda)_2^{<\omega}$ holds.*

PROOF. Let $\mathcal{A} = \langle \kappa, \ldots \rangle$ be an arbitrary structure in a countable language and $\theta < \kappa$ be arbitrary. Let $g : [\theta]^{<\omega} \to 2$ be a function in the ground model that has no homogeneous sets (in the ground model) of ordertype λ, and consider the structure

$$\bar{\mathcal{A}} = \mathcal{A}^\frown \langle \theta, g \restriction [\theta]^n \rangle_{n \in \omega},$$

where θ, and each $g \restriction [\theta]^n$ is considered as a relation. Let $\{\varphi_n \ ; n < \omega\}$ enumerate the formulas of the language of $\bar{\mathcal{A}}$ so that each φ_n has $k(n) < n$ many free variables. Define $f : [\kappa]^{<\omega} \to 2$

by $f(\xi_1,\ldots,\xi_n) = 1$ iff $\mathcal{A} \models \varphi_n(\xi_1,\ldots,\xi_{k(n)})$ and $f(\xi_1,\ldots,\xi_n) = 0$ otherwise. We call this f the function that describes truth in $\mathcal{A}$. Let $\dot{f}$ be a $\mathbb{P}$-name for f. Since κ is inaccessible in V, $|\mathcal{P}(\mathbb{P})| < \kappa$ in V. In V define the function $h : [\kappa]^{<\omega} \to \mathcal{P}(\mathbb{P})$ by

$$h(x) \stackrel{\text{def}}{=} \{p \in \mathbb{P} \,;\, p \Vdash \dot{f}(\check{x}) = 0\}.$$

By Lemma 0.15 let $A \in [\kappa]^\lambda$ be homogeneous for h. Note that since we have attached g in $\mathcal{A}$, $A \subseteq \kappa \setminus \theta$. We will show that A is homogeneous for f in $V[G]$.

Let $n \in \omega$ and $x \in [A]^n$ be arbitrary.

- If $h(x) = \varnothing$ then for all $p \in \mathbb{P}$, $p \nVdash \dot{f}(\check{x}) = \check{0}$. So for some $p \in G \cap E_\gamma$, $p \Vdash \dot{f}(\check{x}) = \check{1}$ and so the colour of $[A]^n$ is 1.
- If $h(x) \neq \varnothing$ and $h(x) \cap G \neq \varnothing$ then the colour of $[A]^n$ is 0.
- If $h(x) \neq \varnothing$ and $h(x) \cap G = \varnothing$ then assume for a contradiction that for some $y \in [A]^n$, $f(x) \neq f(y)$. Without loss of generality say $f(y) = 0$. But then there is $p \in G$ such that $p \Vdash \dot{f}(\check{y}) = \check{0}$ so $\varnothing \neq h(y) \cap G = h(x) \cap G$, contradiction.

So in $V[G]$, $\kappa \to^\theta (\lambda)_2^{<\omega}$ holds. qed

We can use this to get the following.

LEMMA 3.46. *If V is a model of ZFC+ "$\kappa = \kappa(\lambda)$ exists", then for any regular cardinal $\eta < \kappa$, there is a symmetric model $V(G)$ of ZF in which for every $\theta < \kappa$, $\eta^+ \to^\theta (\lambda)_2^{<\omega}$ holds.*

PROOF. Let $\eta < \kappa$ be a regular cardinal, and construct the Jech model $V(G)$ in Section 3 of Chapter 1 that makes $\kappa = \eta^+$. The approximation lemma holds in this model. Let $\theta < \kappa$ be arbitrary. Let $\mathcal{A} = \langle \kappa, \ldots \rangle$ be an arbitrary first order structure with a countable language and let $\dot{\mathcal{A}} \in \mathsf{HS}$ be a name for $\mathcal{A}$ with support E_γ for some $\eta < \gamma < \kappa$. By the approximation lemma

$$\mathcal{A} \in V[G \cap E_\gamma].$$

Note that $|E_\gamma| < \kappa$ therefore by Lemma 3.45, in $V(G)$, $\kappa \to^\theta (\lambda)_2^{<\omega}$ holds. Therefore the structure $\mathcal{A}$ has a set of indiscernibles $A \in [\kappa \setminus \theta]^\lambda$ and $A \in V[G \cap E_\gamma] \subseteq V(G)$. qed

By Corollary 3.8 we get the following.

COROLLARY 3.47. *If V is a model of ZFC with a cardinal κ that is the λ-Erdős cardinal then for any $\eta < \kappa$ regular cardinal there is a symmetric model $V(G)$ in which for every $\theta < \kappa$ and $\rho < \lambda$*

$$(\eta^+, \theta) \twoheadrightarrow (\lambda, \rho) \textit{ holds.}$$

Note that as with many of our forcing constructions here, this η could be *any* predefined regular ordinal of V. So we get an infinity of consistency strength results, some of them looking very strange for someone accustomed to the theory ZFC, such as the following.

COROLLARY 3.48. *If $V \models$ ZFC+ "$\kappa(\omega_3)$ exists" then there is a symmetric model $V(G) \models$ ZF $+ (\omega_{23}, \omega_{17}) \twoheadrightarrow (\omega_3, \omega_2)$.*

Or even stranger:

Corollary 3.49. *If* $V \models \mathsf{ZFC}+$ *"*$\kappa(\omega_\omega)$ *exists" then there is a symmetric model* $V(G) \models \mathsf{ZF} + (\omega_{\omega+3}, \omega_\omega) \twoheadrightarrow (\omega_\omega, \omega_2)$.

To get Chang conjectures that involve more than four cardinals we will have to collapse the Erdős cardinals simultaneously. We will give an example in which the Chang conjecture

$$(\omega_4, \omega_2, \omega_1) \twoheadrightarrow (\omega_3, \omega_1, \omega)$$

holds. Before we do that let us get a very useful proposition.

Proposition 3.50. *Assume that* $V \models \mathsf{ZFC}+$ *"*$\kappa = \kappa(\lambda)$ *exists",* $\mathbb{P}$ *is a partial order such that* $|\mathbb{P}| < \kappa$, *and* $\mathbb{Q}$ *is a partial order that doesn't add subsets to* κ. *If* G *is* $\mathbb{P} \times \mathbb{Q}$*-generic then for every* $\theta < \kappa$,

$$V[G] \models \kappa \to^\theta (\lambda)_2^{<\omega}.$$

Proof. Let $\mathcal{A} = \langle \kappa, \ldots \rangle$ be an arbitrary structure with a countable language, in $V[G]$. By Proposition 1.11, $G = G_1 \times G_2$ for some G_1 $\mathbb{P}$-generic and some G_2 $\mathbb{Q}$-generic. Since $\mathbb{Q}$ does not add subsets to κ, we have that $\mathcal{A} \in V[G_1]$. By Lemma 3.45 we get that $\kappa \to^\theta (\lambda)_2^{<\omega}$ in $V[G_1] \subset V[G]$ and from that we get a set $H \in [\kappa \setminus \theta]^\lambda$ of indiscernibles for $\mathcal{A}$ with respect to parameters below θ, and $H \in V[G]$,. Therefore $V[G] \models \kappa \to^\theta (\lambda)_2^{<\omega}$. qed

We will construct a symmetric model similar to the in the same way we constructed the model with alternating measurables in Section 4 of Chapter 1.

Lemma 3.51. (ZFC) *Assume that* $\kappa_1 = \kappa(\omega_1)$, *and* $\kappa_2 = \kappa(\kappa_1^+)$ *exist. Then there is a symmetric extension of* V *in which* $\mathsf{ZF} + \omega_4 \to^{\omega_2} (\omega_3)_2^{<\omega} + \omega_2 \to^{\omega_1} (\omega_1)_2^{<\omega}$.

Consequently,

$$(\omega_4, \omega_2, \omega_1) \twoheadrightarrow (\omega_3, \omega_1, \omega)$$

holds in V *as well.*

Proof. We will use the model in Section 4 of Chapter 1 for $\rho = 2$. We include the construction here so that our argument is self contained. This construction can be illustrated as below.

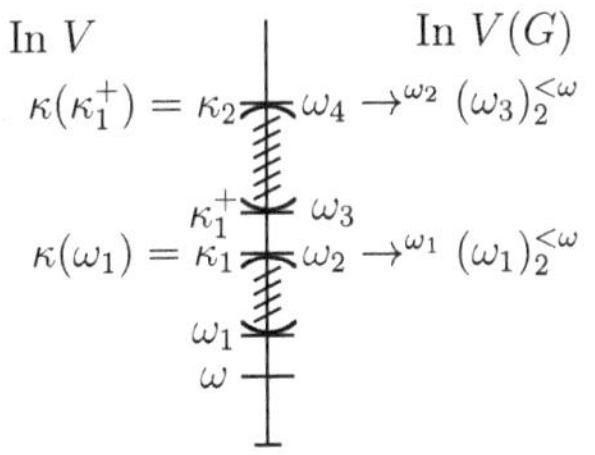

Let $\kappa_1' = (\kappa_1^+)^V$ and define

$$\mathbb{P} \stackrel{\text{def}}{=} \mathsf{Fn}(\omega_1, \kappa_1, \omega_1) \times \mathsf{Fn}(\kappa_1', \kappa_2, \kappa_1').$$

Let $\mathcal{G}_1$ be the full permutation group of κ_1 and $\mathcal{G}_2$ the full permutation group of κ_2. We define an automorphism group $\mathcal{G}$ of $\mathbb{P}$ by letting $a \in \mathcal{G}$ iff for some $a_1 \in \mathcal{G}_1$ and $a_2 \in \mathcal{G}_2$,

$$a((p_1, p_2)) \stackrel{\text{def}}{=} (\{(\xi_1, a_1(\beta_1)) \,;\, (\xi_1, \beta_1) \in p_1\}, \{(\xi_2, a_2(\beta_2)) \,;\, (\xi_2, \beta_2) \in p_2\}).$$

Let I be the symmetry generator that is induced by the ordinals in the product of intervals $(\omega_1, \kappa_1) \times (\kappa_1', \kappa_2)$, i.e.,

$$I \stackrel{\text{def}}{=} \{E_{\alpha,\beta} \,;\, \alpha \in (\omega_1, \kappa_1) \text{ and } \beta \in (\kappa_1', \kappa_2)\},$$

where

$$E_{\alpha,\beta} \stackrel{\text{def}}{=} \{(p_1 \cap (\omega_1 \times \alpha), p_2 \cap (\kappa_1' \times \beta)) \,;\, (p_1, p_2) \in \mathbb{P}\}.$$

This I is clearly a projectable symmetry generator with projections

$$(p_1, p_2)\restriction^* E_{\alpha,\beta} = (p_1 \cap (\omega_1 \times \alpha), p_2 \cap (\kappa_1' \times \beta)).$$

Take the symmetric model $V(G) = V(G)^{\mathcal{F}_I}$. It's easy to see that the approximation lemma holds for this model.

With the standard arguments we can show that in $V(G)$ we have that $\kappa_1 = \omega_2$ and $\kappa_2 = \omega_4$. We want to show that moreover $\kappa_2 \to^{\kappa_1} (\kappa_1')^{<\omega}_2$ and $\kappa_1 \to^{\omega_1} (\omega_1)^{<\omega}_2$.

For the first partition property let $\mathcal{A} = \langle \kappa_2, \ldots \rangle$ be an arbitrary structure in a countable language and let $\dot{\mathcal{A}} \in \mathsf{HS}$ be a name for $\mathcal{A}$ with support $E_{\alpha,\beta}$. By the approximation lemma we have that $\mathcal{A} \in V[G \cap E_{\alpha,\beta}]$. Since $|E_{\alpha,\beta}| < \kappa_2$, by Lemma 3.45 we have that $V[G \cap E_{\alpha,\beta}] \models \kappa_2 \to^{\kappa_1} (\kappa_1')^{<\omega}_2$ therefore there is a set $A \in [\kappa_2 \setminus \kappa_1]^{\kappa_1'}$ of indiscernibles for $\mathcal{A}$ with respect to parameters below κ_1, and $A \in V[G \cap E_{\alpha,\beta}] \subseteq V(G)$.

For the second partition property let $\mathcal{B} = \langle \kappa_1, \ldots \rangle$ be and arbitrary structure in a countable language and let $\dot{\mathcal{B}} \in \mathsf{HS}$ be a name for $\mathcal{B}$ with support $E_{\gamma,\delta}$. We have that $E_{\gamma,\delta} = \mathsf{Fn}(\omega_1, \gamma, \omega_1) \times \mathsf{Fn}(\kappa_1', \delta, \kappa_1')$, $|\mathsf{Fn}(\omega_1, \gamma, \omega_1)| < \kappa_1$, and $\mathsf{Fn}(\kappa_1', \delta, \kappa_1')$ does not add subsets to κ_1. Therefore by Proposition 3.50 we get that $V[G \cap E_{\gamma,\delta}] \models \kappa_1 \to^{\omega_1} (\omega_1)^{<\omega}_2$ so there is a set $B \in [\kappa_1 \setminus \omega_1]^{\omega_1}$ of indiscernibles for $\mathcal{B}$ with respect to parameters below ω_1, and $B \in V[G \cap E_{\gamma,\delta}] \subseteq V(G)$.

So in $V(G)$ we have that $\omega_4 \to^{\omega_2} (\omega_3)^{<\omega}_2$ and $\omega_2 \to^{\omega_1} (\omega_1)^{<\omega}_2$ thus by Lemma 3.9 we have that in $V(G)$ the Chang conjecture $(\omega_4, \omega_2, \omega_1) \twoheadrightarrow (\omega_3, \omega_1, \omega)$ holds. qed

Note that, as discussed, the gap in these cardinals is necessary for this method to work. Collapsing further would destroy their large cardinal properties. Keeping this in mind it is easy to see how to modify this proof to get any desired Chang conjecture

$$(\kappa_n, \ldots, \kappa_0) \twoheadrightarrow (\lambda_n, \ldots, \lambda_0)$$

with the κ_i and/or the λ_i are any predefined successor cardinals, as long as we mind the gaps.

We can do this for the infinitary version as well, using a finite support product forcing of such collapses, for a coherent sequence of Erdős cardinals $\langle \kappa_n \,;\, n \in \omega \rangle$ with respect to

$\langle \kappa_n^+ \,;\, n < \omega \rangle$, and with $\kappa_0 = \omega_1$. In that case we would end up with a model of

$$\mathsf{ZF} + \neg \mathsf{AC}_\omega + (\omega_{2n+1})_{n<\omega} \twoheadrightarrow (\omega_{2n})_{n<\omega}.$$

LEMMA 3.52. *Let $\langle \kappa_n \,;\, n < \omega \rangle$ and $\langle \lambda_n \,;\, 0 < n < \omega \rangle$ be increasing sequences of cardinals such that $\langle \kappa_n \,;\, n < \omega \rangle$ is a coherent sequence of Erdős cardinals with respect to $\langle \lambda_n \,;\, n < \omega \rangle$. If $\mathbb{P}$ is a partial order of cardinality $< \kappa_1$ and G is $\mathbb{P}$-generic then in $V[G]$, $\langle \kappa_n \,;\, n < \omega \rangle$ is a coherent sequence of cardinals with the property $\kappa_{n+1} \to^{\kappa_n} (\lambda_{n+1})_2^{<\omega}$.*

PROOF. Let $\kappa = \bigcup_{n\in\omega}$ and let $\mathcal{A} = \langle \kappa, \ldots \rangle \in V[G]$ be an arbitrary structure in a countable language. Let $\{\varphi_n \,;\, n < \omega\}$ enumerate the formulas of the language of $\mathcal{A}$ so that each φ_n has $k(n) < n$ many free variables. Define $f : [\kappa]^{<\omega} \to 2$ by $f(\xi_1, \ldots, \xi_n) = 1$ iff $\mathcal{A} \models \varphi_n(\xi_1, \ldots, \xi_{k(n)})$ and $f(\xi_1, \ldots, \xi_n) = 0$ otherwise. Let $\dot{f}$ be a $\mathbb{P}$-name for f. In V define a function $g : [\kappa]^{<\omega} \to \mathcal{P}(\mathbb{P})$ by

$$g(x) = \{p \in \mathbb{P} \,;\, p \Vdash \dot{f}(\check{x}) = \check{0}\}.$$

Since $|\mathbb{P}| < \kappa_1$ and κ_1 is inaccessible in V, $|\mathcal{P}(\mathbb{P})| < \kappa_1$. So there is a $\langle \lambda_n \,;\, 0 < n < \omega \rangle$-coherent sequence of homogeneous sets for g with respect to $\langle \kappa_n \,;\, n \in \omega \rangle$. The standard arguments show that this is a $\langle \lambda_n \,;\, 0 < n < \omega \rangle$-coherent sequence of homogeneous sets for f with respect to $\langle \kappa_n \,;\, n \in \omega \rangle$, therefore a $\langle \lambda_n \,;\, 0 < n < \omega \rangle$-coherent sequence of indiscernibles for $\mathcal{A}$ with respect to $\langle \kappa_n \,;\, n \in \omega \rangle$. qed

The model used for this following proof is the same one we build for the alternating measurables in Section 4 of Chapter 1.

LEMMA 3.53. (ZFC) *Let $\langle \kappa_n \,;\, n \in \omega \rangle$ be a coherent sequence of Erdős cardinals with respect to $\langle \lambda_n \,;\, 0 < n \in \omega \rangle$, where $\kappa_0 = \omega_1$. Then there is a symmetric model $V(G)$ in which $\langle \omega_{2n} \,;\, n \in \omega \rangle$ is a coherent sequence of cardinals with the property $\omega_{2n+2} \to^{\omega_{2n}} (\omega_{2n+1})_2^{<\omega}$.*

Consequently, in $V(G)$

$$(\ldots, \omega_{2n}, \ldots, \omega_4, \omega_2, \omega_1) \twoheadrightarrow (\ldots, \omega_{2n-1}, \ldots, \omega_3, \omega_1, \omega)$$

holds as well.

PROOF. We will construct the model in Section 4 of Chapter 1 for $\rho = \omega$. In order for the proof to be clear, we will repeat the definition of that model here.

Let $\kappa = \bigcup_{0<n<\omega} \kappa_n$, for every $0 < n < \omega$ let $\kappa_n' = \kappa_n^+$, and let $\kappa_0' = \omega_1$. For every $0 < n < \omega$ let

$$\mathbb{P}_n \stackrel{\text{def}}{=} \mathsf{Fn}(\kappa_{n-1}', \kappa_n, \kappa_{n-1}'),$$

and take the finite support product of these forcings

$$\mathbb{P} \stackrel{\text{def}}{=} \prod_{0<n<\omega}^{\text{fin}} \mathbb{P}_n.$$

For each $0 < n < \omega$ let G_n be the full permutation group of κ_n and define an automorphism group $\mathcal{G}$ of $\mathbb{P}$ by $a \in \mathcal{G}$ iff for every $n \in \omega$ there exists $a_n \in \mathcal{G}_n$ such that

$$a(\langle p_n \,;\, n \in \omega \rangle) \stackrel{\text{def}}{=} \langle \{(\xi, a_n(\beta)) \,;\, (\xi, \beta) \in p_n\} \,;\, n \in \omega \rangle.$$

For every finite sequence of ordinals $e = \langle \alpha_1, \dots, \alpha_m \rangle$ such that for every $i = 1, \dots, m$ there is a distinct $0 < n_i < \omega$ such that $\alpha_i \in (\kappa'_{n_i-1}, \kappa_{n_i})$, define

$$E_e \stackrel{\text{def}}{=} \{\langle p_{n_i} \cap (\kappa'_{n_i-1}, \alpha_i) \; ; \; \alpha_i \in e \rangle \; ; \; \langle p_{n_i} \; ; \; i = 1, \dots, m \rangle \in \mathbb{P}\},$$

and take the symmetry generator

$$I \stackrel{\text{def}}{=} \{E_e \; ; \; e \in \prod_{0<n<\omega}^{\text{fin}} (\kappa'_{n-1}, \kappa_n)\}.$$

This is clearly a projectable symmetry generator with projections

$$\langle p_j \; ; \; 0 < j < \omega \rangle {\upharpoonright}^* E_e = \langle p_{n_i} \cap (\kappa'_{n_i-1}, \alpha_i) \; ; \; \alpha_i \in e \rangle.$$

Take the symmetric model $V(G) = V(G)^{\mathcal{F}_I}$. It's clear that the approximation lemma holds for $V(G)$.

As usual we can show that in $V(G)$, for each $0 < n < \omega$ we have that $\kappa_n = \kappa'^{+}_{n-1}$, i.e., for every $0 < n < \omega$, $\kappa_n = \omega_{2n}$ and $\kappa'_n = \omega_{2n+1}$.

It remains to show that $\langle \kappa_i \; ; \; i \in \omega \rangle$ is a coherent sequence of cardinals with the property $\kappa_{n+1} \to^{\kappa_n} (\lambda_{n+1})^{<\omega}_2$.

Let $\mathcal{A} = \langle \kappa, \dots \rangle$ be an arbitrary structure in a countable language and let the function $f : [\kappa]^{<\omega} \to 2$ describe the truth in $\mathcal{A}$, as in the proofs of Lemma 3.45 and Lemma 3.52. Let $\dot{f} \in \mathsf{HS}$ be a name for f with support E_e. Let $e = \{\alpha_1, \dots, \alpha_m\}$ and for each $i = 1, \dots, m$ let n_i be such that $\alpha \in (\kappa'_{n_i-1}, \kappa_{n_i})$. By the approximation lemma,

$$f \in V[G \cap E_e],$$

i.e., f is forced via $\bar{\mathbb{P}} = \prod_{i=1}^m \mathsf{Fn}(\kappa'_{n_i-1}, \kappa_{n_i}, \kappa'_{n_i-1})$.

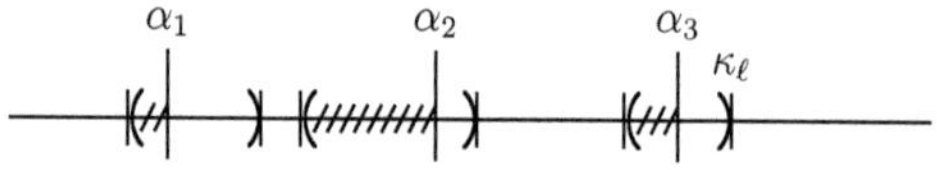

Let $\ell \stackrel{\text{def}}{=} \max\{n_i \; ; \; \alpha_i \in e\}$. We're in a situation as in the image above, which is an example for $m = 3$. Since $|\mathbb{P}| < \kappa_\ell$, by Lemma 3.52 there is a $\langle \lambda_n \; ; \; \ell \leq n < \omega \rangle$-coherent sequence of indiscernibles for $\mathcal{A}$ with respect to $\langle \kappa_n \; ; \; \ell - 1 \leq n \in \omega \rangle$, i.e., a sequence $\langle A_n \; ; \; \ell \leq n < \omega \rangle$ such that

- for every $\ell \leq n < \omega$, $A_n \subseteq \kappa_n \setminus \kappa_{n-1}$ is of ordertype λ_n, and
- if $x, y \in [\kappa]^{<\omega}$ are such that $x = \{x_1, \dots, x_m\}$, $y = \{y_1, \dots, y_m\}$, $x, y \in \bigcup_{\ell \leq n < \omega} A_n$, and for every $\ell \leq n < \omega$, $|x \cap A_n| = |y \cap A_n|$, then for every $m + k$-ary formula φ in the language of $\mathcal{A}$, and every $z_1, \dots, z_k$ less than $\min \bigcup_{\ell \leq n < \omega} A_n$,

$$\mathcal{A} \models \varphi(z_1, \dots, z_k, x_1, \dots, x_m) \iff \mathcal{A} \models \varphi(z_1, \dots, z_k, y_1, \dots, y_m)$$

Now we will get sets of indiscernibles from the remaining cardinals $\kappa_1, \ldots, \kappa_{\ell-1}$ step by step, making them coherent as we go along. Before we get the rest of the A_n, note that by Proposition 3.50 we have that for every $0 < n < \ell$,

$$V[G \cap E_e] \models \kappa_n \rightarrow^{\kappa_{n-1}} (\lambda_n)_2^{<\omega}.$$

Let's see how to get $A_{\ell-1}$. For every $\ell \leq n < \omega$, let $\bar{A}_n$ be the first ω-many elements of A_n. There are only countably many $x \in [\kappa]^{<\omega}$ such that $x \subseteq \bigcup_{\ell \leq n < \omega} \bar{A}_n$. For every $i, j \in \omega$, and every $x \in [\kappa]^{<\omega}$ such that $x = \{x_1, \ldots, x_m\} \subseteq \bigcup_{\ell \leq n < \omega} \bar{A}_n$ and $m < i, j$, let

$$f_{i,x}(v_1, \ldots, v_{i-m}) \stackrel{\text{def}}{=} f_i(v_1, \ldots, v_{i-m}, x_1, \ldots, x_m), \text{ and}$$
$$R_{j,x}(v_1, \ldots, v_{j-m}) \stackrel{\text{def}}{=} R_j(v_1, \ldots, v_{j-m}, x_1, \ldots, x_m).$$

Consider the structure

$$\mathcal{A}' \stackrel{\text{def}}{=} \mathcal{A}^\frown \langle f_{i,x}, R_{j,x} \rangle_{i,j<\omega, x \in [\kappa]^{<\omega}, x = \{x_1,\ldots,x_m\} \subseteq \bigcup_{\ell \leq n < \omega} \bar{A}_n, m < i,j}.$$

Since $\kappa_{\ell-1} \rightarrow^{\kappa_{\ell-2}} (\lambda_{\ell-1})_2^{<\omega}$, there is a set $A_{\ell-1} \in [\kappa_{\ell-1} \setminus \kappa_{\ell-2}]^{\lambda_{\ell-1}}$ of indiscernibles for $\mathcal{A}'$ with respect to parameters below $\kappa_{\ell-2}$. By the way we defined $\mathcal{A}'$, the sequence $\langle A_n \,;\, \ell-1 \leq n < \omega \rangle$ is a $\langle \lambda_n \,;\, \ell-1 \leq n < \omega \rangle$-coherent sequence of indiscernibles for $\mathcal{A}$ with respect to $\langle \kappa_n \,;\, \ell-2 \leq n < \omega \rangle$.

Continuing in this manner we get a sequence $\langle A_n \,;\, 0 < n < \omega \rangle$ that is a $\langle \lambda_n \,;\, 0 < n < \omega \rangle$-coherent sequence of indiscernibles for $\mathcal{A}$ with respect to $\langle \kappa_n \,;\, n < \omega \rangle$, and such that $\langle A_n \,;\, 0 < n < \omega \rangle \in V[G \cap E_e] \subseteq V(G)$.

Therefore we have that in $V(G)$, $\langle \omega_{2n} \,;\, n \in \omega \rangle$ is a coherent sequence of cardinals with the property $\omega_{2n+2} \rightarrow^{\omega_{2n}} (\omega_{2n+1})_2^{<\omega}$. By Corollary 3.15 we have that in $V(G)$

$$(\ldots, \omega_{2n}, \ldots, \omega_4, \omega_2, \omega_1) \twoheadrightarrow (\ldots, \omega_{2n-1}, \ldots, \omega_3, \omega_1, \omega)$$

holds. qed

Note that in this model the axiom of choice fails badly. In particular, by Lemma 1.37 $\mathsf{AC}_{\omega_2}(\mathcal{P}(\omega_1))$ is false. As we mentioned after the proof of that lemma, one can get the infinitary Chang conjecture plus the axiom of dependent choice with the construction in the proof of [**AK06**, Theorem 5]. With that we get the following.

Lemma 3.54. *Let $V_0 \models$ ZFC+ "there exists a measurable cardinal κ". Let $n < \omega$ be fixed but arbitrary. There is a generic extension V of V_0, a forcing notion $\mathbb{P}$, and a symmetric model N such that*

$$N \models \mathsf{ZF} + \mathsf{DC} + (\omega_n)_{0<n<\omega} \twoheadrightarrow (\omega_n)_{n<\omega}.$$

3.1.1. *Weak Chang conjecture.* To get $\mathsf{wCc}(\tau^+, \tau)$ from the existence of the almost $< \tau$-Erdős, we will construct a model where τ^+ is the almost $< \tau$-Erdős, using the generalised Jech construction.

Lemma 3.55. *If V is a model of* ZFC + *"there exist regular cardinals $\kappa > \tau$ such that κ is the almost $< \tau$-Erdős cardinal, then there is a symmetric extension of V which is a model of* $\mathsf{ZF} + \neg\mathsf{AC} + \tau^+ \rightarrow^\tau (< \tau)_2^{<\omega}$.

PROOF. Construct the Jech symmetric model $V(G)$ as shown in Section 3 of Chapter 1, by symmetrically collapsing κ to become τ^+. Let $\mathcal{A} = \langle \kappa, \ldots \rangle$ be an arbitrary structure in a countable language and $h : [\tau]^{<\omega} \to 2$ a function in the ground model that has no homogeneous sequences of order τ. Let $\bar{\mathcal{A}} \stackrel{\text{def}}{=} \mathcal{A}^\frown \langle h\restriction[\tau]^n \rangle_{n\in\omega}$ and define $f : [\kappa]^{<\omega} \to 2$ to describe the truth in $\bar{\mathcal{A}}$ as usual. Let $\dot{f} \in \mathsf{HS}$ be a name for f with support E_δ for some $\delta \in (\tau, \kappa)$. By the approximation lemma, $f \in V[G \cap E_\delta]$.

In V, define the function $g : [\kappa]_2^{<\omega} \to \mathcal{P}(E_\delta)$ by

$$g(x) \stackrel{\text{def}}{=} \{p \in E_\delta\ ;\ p \Vdash \dot{f}(\check{x}) = 0\}.$$

Note that $|E_\delta| < \kappa$ and since κ is inaccessible in V, $|\mathcal{P}(E_\delta)| < \kappa$ in V. So by Lemma 3.19, there exists $\langle H_\beta\ ;\ \beta < \tau \rangle$ homogeneous sequence for g. Note that because we attached h to our construction, for every $\beta < \tau$, $H_\beta \subseteq \kappa \setminus \tau$. With the same arguments as in the proof of Lemma 3.45 we get that $\langle H_\beta\ ;\ \beta < \tau \rangle$ is a homogeneous sequence of order τ for f in $V[G \cap E_\delta] \subseteq V(G)$. Thus $\langle H_\beta\ ;\ \beta < \tau \rangle$ is such a sequence.

If $c = \varnothing$. Let $x \in [H_\beta]^n$ be arbitrary. Then for every $q \in \mathsf{rng}(\ddot{f})$ we have $q \nVdash \ddot{f}(\check{x}) = \check{0}$. There is a $q \in \mathsf{rng}(\ddot{f})$ that decides the value of $\ddot{f}(\check{x})$ and for every $q \in \mathsf{rng}(\ddot{f})$ if q decides the value of $\ddot{f}(\check{x})$ then $q \Vdash \ddot{f}(\check{x}) = \check{1}$. Since $\mathsf{rng}(\ddot{f})$ is a dense set, there is a $p \in G \cap E_\alpha$ such that $p \Vdash \ddot{f}(\check{x}) = \check{1}$. So in $V(G)$, $f``[H_\beta]^n = \{1\}$.

If $c \neq \varnothing$. Let $x \in [H_\beta]^n$ be arbitrary. Then there is a $q \in \mathsf{rng}(\ddot{f})$ such that $q \Vdash \ddot{f}(\check{x}) = \check{0}$. If $c \cap G \cap E_\alpha \neq \varnothing$ then $f``[H_\beta]^n = \{0\}$. If $c \cap G \cap E_\alpha = \varnothing$ then assume for a contradiction that $f(x) = 0$ in $V[G \cap E_\alpha]$. There must be some $p \in G \cap E_\alpha$ such that $p \Vdash \ddot{f}(\check{x}) = \check{0}$. But then $p \in c$, contradiction.

The ordertype requirement and that the have the same type with respect to f follows because this is a homogeneous sequence for g. So f has a homogeneous sequence, i.e., κ is almost $< \tau$-Erdős in $V(G)$, and $\kappa = \tau^+$. qed

Therefore by Corollary 3.24 we get the following.

COROLLARY 3.56. *If V is a model of* ZFC+ *"$\tau < \kappa$ is a regular cardinal and κ is the almost $< \tau$-Erdős" then there is a symmetric extension $V(G)$ in which* $\mathsf{wCc}(\tau^+, \tau)$ *holds.*

3.2. Getting indiscernibles. In this subsection we will work with the Dodd-Jensen core model to get strength from the principles we're looking at, by using the lemmas listed in Section 2. We will start by looking at two cardinal Chang conjectures.

THEOREM 3.57. *Assume* ZF *and let η be a regular cardinal. If for some infinite cardinals θ, λ, and ρ such that $\eta^+ > \theta, \lambda > \rho$ and $\mathsf{cf}\lambda > \omega$ the Chang conjecture $(\eta^+, \theta) \twoheadrightarrow (\lambda, \rho)$ holds, then $\kappa(\lambda)$ exists in the Dodd-Jensen core model $(K^{\mathsf{DJ}})^{\mathsf{HOD}}$, and $(K^{\mathsf{DJ}})^{\mathsf{HOD}} \models (\eta^+)^V \to (\lambda)_2^{<\omega}$.*

PROOF. Let $\kappa = (\eta^+)^V$ and in K^{DJ} let $g : [\kappa]^{<\omega} \to 2$ be arbitrary and consider the structure $\langle K^{\mathsf{DJ}}_\kappa, \in, D \cap K^{\mathsf{DJ}}_\kappa, g \rangle$. We want to find a good set of indiscernibles for this structure in K^{DJ}. Using our Chang conjecture in V we get an elementary substructure

$$\mathcal{K}' = \langle K', \in, D \cap K', g' \rangle \prec \langle K^{\mathsf{DJ}}_\kappa, \ldots \rangle$$

such that $|K'| = \lambda$ and $|K' \cap \theta| = \rho$. Since K' is wellorderable it can be seen as a set of ordinals. We attach K' to HOD, getting $\mathsf{HOD}[K']$. By Lemma 3.44, $(K^{\mathsf{DJ}})^{\mathsf{HOD}} = (K^{\mathsf{DJ}})^{\mathsf{HOD}[K']}$.

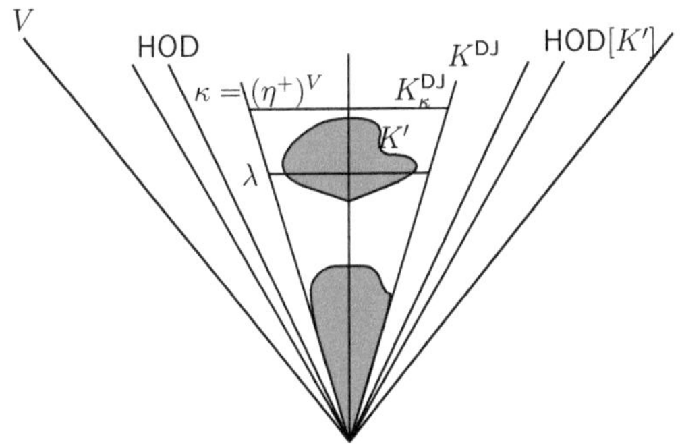

We are now, and for the rest of this proof, working in $\mathsf{HOD}[K']$.

Let $\langle \bar{K}, \in, A' \rangle$ be the Mostowski collapse of $\mathcal{K}'$, with $\pi\colon \bar{K} \to K'$ being an elementary embedding.

We distinguish two cases.

Case 1. If $\bar{K} = K^{\mathsf{DJ}}_{\lambda'}$ for some λ'. Then the map $\pi : K^{\mathsf{DJ}}_{\lambda'} \to K^{\mathsf{DJ}}_{\kappa}$ is elementary.

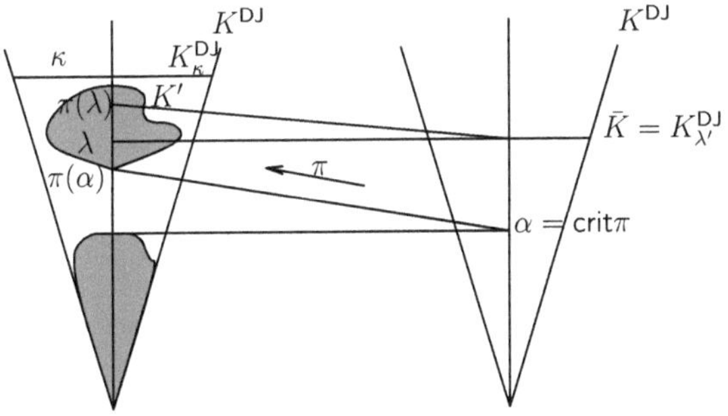

Since $\lambda \geq \omega_1$, by Lemma 3.33, there is a non trivial elementary embedding of K^{DJ} to K^{DJ} with critical point α. By Lemma 3.35 this means that there is an inner model with a measurable cardinal β, such that if $\alpha < \omega_1$ then $\beta \leq \omega_1$ and if $\alpha \geq \omega_1$ then $\beta < \alpha^+$. Because $\alpha = \mathsf{crit}\pi$ and $|\bar{K}| = \lambda$, $\alpha < \lambda^+$.

Let's take a closer look at this inner model. Let U be a normal measure for β in the inner model M, define $\bar{U} \stackrel{\text{def}}{=} U \cap L[U]$ and build $L[\bar{U}]$. It is known that then $L[\bar{U}] \models$"$\bar{U}$ is a normal ultrafilter over β" (Proposition 3.36). We also have that $L[\bar{U}] \models \mathsf{ZFC}$ which by Lemma 3.38 means that $\langle L[\bar{U}], \in, \bar{U} \rangle$ is iterable. Recall that such a structure $\langle L[\bar{U}], \in, \bar{U} \rangle$ is called a β-model and that for a regular cardinal ν, C_ν is the club filter over ν. Lemma 3.39 says that if there is a β-model, if ν is a regular cardinal above β^+, and if $\bar{C}_\nu = C_\nu \cap L[C_\nu]$, then $L[\bar{C}_\nu]$ is a ν-model.

Now, we have that if $\alpha < \omega$ then $\beta \leq \omega_1$, so $\beta \leq \lambda$. If $\alpha \geq \omega_1$ then $\beta < \alpha^+ \leq \lambda^+$, so again $\beta \leq \lambda$. If $\beta = \lambda$ then $L[\bar{U}] \models$"λ is Ramsey". By Lemma 3.37, $\mathcal{P}(\lambda) \cap L[\bar{U}] = \mathcal{P}(\lambda) \cap K^{\mathsf{DJ}}$. So λ is Ramsey in K and we're done.

So assume that $\beta < \lambda$. Then $\beta^+ < \kappa$. We need a regular cardinal $\nu > \beta^+$ such that $\lambda \leq \nu \leq \kappa$. If κ is regular, let $\nu = \kappa$. If κ is singular then κ is a limit cardinal so there is such a regular cardinal ν (e.g., $\nu = \lambda^{++}$).

Then by Lemma 3.37 we have that $L[\bar{C}_\nu] \models$"ν is Ramsey". Because Lemma 3.37 says that $\mathcal{P}(\nu) \cap L[\bar{C}_\nu] = \mathcal{P}(\nu) \cap K^{\mathsf{DJ}}$, this ν is Ramsey in K^{DJ}. But this implies that in K^{DJ}, $\kappa \to (\lambda)_2^{<\omega}$.

Case 2. If $\bar{K} \neq K_\lambda^{\mathsf{DJ}}$ for any λ'. By Lemma 3.31 $K_\kappa^{\mathsf{DJ}} \models$"$V = K^{\mathsf{DJ}}$". Since $\bar{K}$ is elementary with K_κ^{DJ}, $\bar{K} \models$"$V = K^{\mathsf{DJ}}$". This is because being the lower part of a premouse is a property describable by a formula. Let $x \in \bar{K}$. Since $\bar{K} \models$"$V = K^{\mathsf{DJ}}$", there must be some M such that

$$\bar{K} \models \text{“}M \text{ is an iterable premouse and } x \in \mathsf{lp}(M).$$

So $K_\kappa^{\mathsf{DJ}} \models$ "M is an iterable premouse" and by Lemma 3.27, $\pi(M)$ is an iterable premouse in $\mathsf{HOD}[K']$. Since $\pi \restriction M \to \pi(M)$ is elementary and $\pi(M)$ is an iterable premouse, by Lemma 3.28, M is an iterable premouse in $\mathsf{HOD}[K']$. Thus $x \in K^{\mathsf{DJ}}$, so $\bar{K} \subseteq K^{\mathsf{DJ}}$. But then, since $K_\lambda^{\mathsf{DJ}} \neq \bar{K}$ for any λ', and $\bar{K}$ has cardinality λ, there must be an iterable premouse $M \notin \bar{K}$ and a $z \in K_\lambda^{\mathsf{DJ}} \setminus \bar{K}$ such that $\mathsf{lp}(M) \cap (K_\lambda \setminus \bar{K}) \neq \varnothing$, $z \in \mathsf{lp}(M)$, and $M \in K_\lambda^{\mathsf{DJ}}$. Fix M.

Claim 1.
If $\delta > \lambda$ is a regular cardinal then for every iterable premouse $N \in \bar{K}$, $N_\delta \in M_\delta$.

PROOF OF CLAIM. Since $M \in K_\lambda$ and $|\bar{K}| = \lambda$ by Lemma 3.40 we have that for every regular cardinal $\delta > \lambda$ and every iterable premouse $N \in \bar{K}$, N_δ and M_δ are comparable. Assume for a contradiction that for some $N \in \bar{K}$, $M_\delta \subseteq N_\delta$. Then $z \in \mathsf{lp}(N_\delta)$ and since $z \in K_\lambda^{\mathsf{DJ}}$, for some $\xi < \lambda$, $z \in \mathsf{lp}(N_\xi)$. But since $N_\xi \in \bar{K}$, $z \in N_\xi \in \bar{K}$ which is transitive so $z \in \bar{K}$, contradiction.
qed claim

We want such a $\delta \leq \kappa$. As before, if κ is regular then take $\delta = \kappa$, and if κ is singular then take $\delta = \lambda^+$. Look at M_δ. By Claim 1 we have that $\bar{K} \subseteq M_\delta$. So $g' \in M_\delta$.

Let $\langle M_i, \pi_{ij}, \gamma_i, U_i \rangle_{i \leq j < \delta}$ be the δ-iteration of M. By Lemma 3.41(1) there is some $x \in M$ and $\vec{\rho} \in {}^{<\omega}\{\gamma_i \, ; \, i < \delta\}$ such that $g' = \pi_{0,\delta}(x)(\vec{\rho})$. Let $\mathcal{C} = \{\gamma_i \, ; \, i < \lambda\}$. By Lemma 3.41(2) there is a sequence $\langle k_n \, ; \, n < \omega \rangle \in {}^{\omega}2$ such that for every $n < \omega$, g'“$[\mathcal{C}]^n = \{k_n\}$.

By elementarity, π“$\mathcal{C}$ is a homogeneous set for g in $\mathsf{HOD}[K']$ and π“$\mathcal{C}$ is a good set of indiscernibles for $\langle K_\kappa^{\mathsf{DJ}}, \in, D \cap K_\kappa^{\mathsf{DJ}}, g \rangle$ of ordertype $\mathsf{cf}\lambda \geq \omega_1$. By Jensen's indiscernibility lemma (Lemma 3.42) there is a homogeneous set for g of ordertype λ in K^{DJ}. qed

Therefore we have the following.

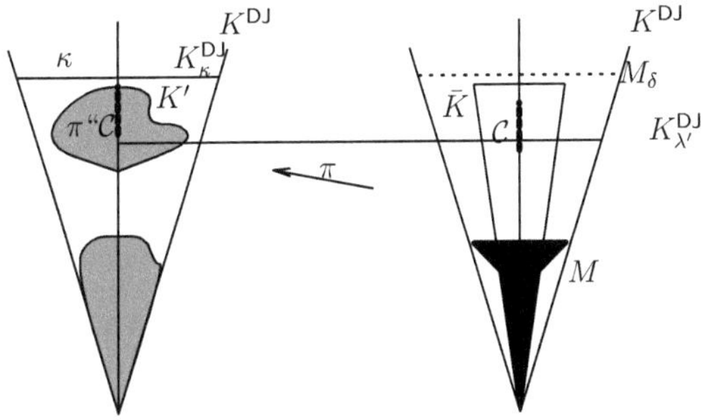

COROLLARY 3.58. *The theory* $\mathsf{ZF} + (\kappa,\theta) \twoheadrightarrow (\lambda,\rho) +$ "$\mathsf{cf}\lambda > \omega$)" *is equiconsistent with the theory* $\mathsf{ZFC}+$ "$\kappa(\lambda)$ *exists*".

In the proof of Lemma 3.9 we see how to combine finitely many sets of indiscernibles to make them coherent. Using this we get the following.

COROLLARY 3.59. *Assume* ZF *and let* $\kappa_n > \cdots > \kappa_0$, $\lambda_n > \cdots > \lambda_0$ *be regular cardinals, such that the Chang conjecture* $(\kappa_n,\ldots,\kappa_0) \twoheadrightarrow (\lambda_n,\ldots,\lambda_0)$ *holds, then for each* $i = 1,\ldots,n$, $\kappa(\lambda_i)$ *exists in the Dodd-Jensen core model* $(K^{\mathsf{DJ}})^{\mathsf{HOD}}$ *and* $(K^{\mathsf{DJ}})^{\mathsf{HOD}} \models \forall i = 1,\ldots,n(\kappa_i \to (\lambda_i)^{<\omega}_2)$.

COROLLARY 3.60. *For every finite* n, *the theory* $\mathsf{ZF}+$ "$(\kappa_n,\ldots,\kappa_0) \twoheadrightarrow (\lambda_n,\ldots,\lambda_0)$" *is equiconsistent with the theory* $\mathsf{ZFC}+$ "$\kappa(\lambda_n^{+(n-1)})$ *exists.*", *where* λ_0 *is the last cardinal appearing on the Chang conjecture.*

For getting the existence of the almost $< \tau$-Erdős cardinal from a weak Chang conjecture, we need to get homogeneous sets of the same type.

THEOREM 3.61. *Assume* ZF, *let* τ *be a regular successor cardinal, and* $\kappa > \tau$ *a cardinal such that* $\mathsf{wCc}(\kappa,\tau)$ *holds. Then* $\kappa \to (<\tau)^{<\omega}_2$ *holds in* $(K^{\mathsf{DJ}})^{\mathsf{HOD}}$.

PROOF. This proof is a modification of the proof of [**DK83**, Theorem D].

We work in HOD until indicated otherwise. Let $g : [\kappa]^{<\omega} \to 2$ be an arbitrary function in K^{DJ} and consider the structure

$$\mathcal{A} \stackrel{\text{def}}{=} \langle K^{\mathsf{DJ}}_\kappa, \in, D, g\rangle,$$

where $D \subseteq \kappa$ is as in the comments after Definition 3.29. If τ is inaccessible in K^{DJ} then get a system $\langle \mathcal{B}_\beta, \sigma_{\gamma\beta} \,;\, \gamma \leq \beta < \tau\rangle$ from $\mathsf{wCc}(\kappa,\tau)$ in V for $\xi = 0$ and if τ is a successor cardinal in K^{DJ} then get $\langle \mathcal{B}_\beta, \sigma_{\gamma\beta} \,;\, \gamma \leq \beta < \tau\rangle$ for $\xi < \tau$ such that $\tau = (\xi^+)^{K^{\mathsf{DJ}}}$. We will show that the second case is impossible as τ will turn out to be inaccessible.

Let $X \subseteq \mathrm{Ord}$ code $\langle \mathcal{B}_\beta, \sigma_{\gamma\beta} \,;\, \gamma \leq \beta < \tau\rangle$ and consider $\mathsf{HOD}[X]$. By Lemma 3.44,

$$(K^{\mathsf{DJ}})^{\mathsf{HOD}} = (K^{\mathsf{DJ}})^{\mathsf{HOD}[X]}.$$

From now on and until the end of the proof we work in $\mathsf{HOD}[X]$, which is a model of ZFC.

For every $\beta < \tau$ let $\pi_\beta : \bar{\mathcal{B}}_\beta \to \mathcal{B}_\beta$ be the Mostowski collapse of $\mathcal{B}_\beta$.

Case 1. *For every iterable premouse M of cardinality less than τ, there is a $\beta < \tau$ such that $\mathsf{Ip}(M) \subseteq \bar{\mathcal{B}}_\beta$.*

We first show that τ is inaccessible in K^{DJ}. Assume for a contradiction that for some ξ, $\tau = (\xi^+)^{K^{\mathsf{DJ}}}$. Note that since for every $\beta, \gamma < \tau$, $\xi \subseteq \mathcal{B}_\beta \cap \tau = \mathcal{B}_\gamma \cap \tau \subseteq \alpha$, we have that there is a $\bar{\alpha} \in (\xi, \tau)$ such that for every $\beta < \tau$, $\pi_\beta(\xi) = \xi$ and $\pi_\beta(\bar{\alpha}) \geq \tau$. Let M be an iterable premouse with $|M| < \tau$ such that $\mathsf{Ip}(M)$ contains a surjective map $f : \xi \to \bar{\alpha}$. By assumption there is some $\beta < \tau$ such that $f \in \mathsf{Ip}(M) \subseteq \bar{\mathcal{B}}_\beta$. But then $\pi_\beta(f)$ is a surjection from ξ onto $\pi_\beta(\bar{\alpha}) \geq \tau$, contradiction to τ being a cardinal.

(4) $$\text{For every } \beta < \tau,\ \bar{\mathcal{B}}_\beta \subseteq K^{\mathsf{DJ}}.$$

To see this let $\beta < \tau$ be arbitrary and let $x \in \bar{\mathcal{B}}_\beta$. Since $\bar{\mathcal{B}}_\beta \models V = K^{\mathsf{DJ}}$, there is some $M \in \bar{\mathcal{B}}_\beta$ such that

$$\bar{\mathcal{B}}_\beta \models \text{“}M \text{ is an iterable premouse and } x \in \mathsf{Ip}(M)\text{''}.$$

So $K^{\mathsf{DJ}}_\kappa \models$ “$\pi_\beta(M)$ is an iterable premouse” and by Lemma 3.27, $\pi_\beta(M)$ is an iterable premouse in $\mathsf{HOD}[X]$. So $\pi_\beta \upharpoonright M : M \to \pi_\beta(M)$ is an elementary map and since $\pi_\beta(M)$ is an iterable premouse we get by Lemma 3.28 that M is an iterable premouse in $\mathsf{HOD}[X]$, hence $x \in K^{\mathsf{DJ}}$.

Now let $\bar{\alpha}$ be the critical point of every π_β, for $\beta < \tau$.

(5) $$\text{For some } \beta < \tau,\ \mathcal{P}(\bar{\alpha}) \cap K^{\mathsf{DJ}} \in \bar{\mathcal{B}}_\beta.$$

To see this let M be an iterable premouse such that $\mathcal{P}(\bar{\alpha}) \cap K^{\mathsf{DJ}} \in \mathsf{Ip}(M)$. Since τ is inaccessible in K^{DJ}, we may assume that $|M| < \tau$ so by assumption there is some $\beta < \tau$ such that $\mathcal{P}(\bar{\alpha}) \cap K^{\mathsf{DJ}} \in \mathsf{Ip}(M) \subseteq \bar{\mathcal{B}}_\beta \subseteq K^{\mathsf{DJ}}$. Therefore, $\mathcal{P}(\bar{\alpha}) \cap K^{\mathsf{DJ}} = (\mathcal{P}(\alpha))^{\bar{\mathcal{B}}_\beta}$ and $\bar{\mathcal{B}}_\beta \models$ “$\mathcal{P}(\bar{\alpha})$ exists”.

Let β be the least such that (5) holds and fix β. It follows by (5) that

$$U \stackrel{\text{def}}{=} \{x \subseteq \bar{\alpha}\,;\, x \in K^{\mathsf{DJ}} \text{ and } \bar{\alpha} \in \pi_0(x)\}$$

is an ultrafilter of $\mathcal{P}(\bar{\alpha}) \cap K^{\mathsf{DJ}}$. It remains to show that $({}^{\bar{\alpha}}K^{\mathsf{DJ}} \cap K^{\mathsf{DJ}})/U$ is well-founded. Assume towards a contradiction that it is not well-founded. Then by Lemma 3.32 there are $f_0, f_1, \cdots \in K^{\mathsf{DJ}}_\tau$ such that for every $i < \omega$,

$$F_i \stackrel{\text{def}}{=} \{\nu < \bar{\alpha}\,;\, f_{i+1}(\nu) \in f_i(\nu)\} \in U.$$

For every $i \in \omega$ let $\gamma_i < \tau$ be such that $f_i \in \bar{\mathcal{B}}_{\gamma_i}$. Since $\tau \geq \omega_1$ is a regular cardinal, there is a $\gamma < \tau$ such that for every $i \in \omega$, $\gamma_i < \gamma$. Then $f_0, f_1, \cdots \in \bar{\mathcal{B}}_\gamma$.

If $\gamma < \beta$ then $\bar{\mathcal{B}}_\gamma \subseteq \bar{\mathcal{B}}_\beta$ because they are transitive. Then $f_0, f_1, \cdots \in \bar{\mathcal{B}}_\beta$ therefore for every $i \in \omega$ we have

$$\bar{\alpha} \in \pi_\beta(\{\nu < \bar{\alpha}\,;\, f_{i+1}(\nu) \in f_i(\nu)\}) = \{\nu < \pi_\beta(\bar{\alpha})\,;\, \pi_\beta(f_{i+1}) \in \pi_\beta(f_i)\}.$$

So $\pi_\beta(f_0)(\bar{\alpha}) \in \pi_\beta(f_1)(\bar{\alpha}) \in \ldots$, contradiction.

If $\beta < \gamma$ then let $\tilde{\sigma}_{\beta\gamma} \stackrel{\text{def}}{=} \pi_\gamma^{-1} \circ \sigma_{\beta\gamma} \circ \pi_\beta$, and note that we have a diagram of embeddings as depicted below.

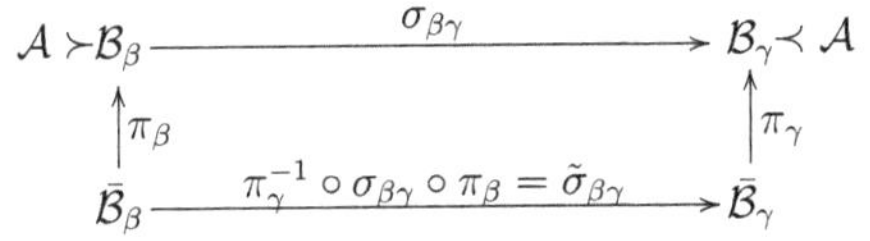

$$\begin{array}{ccc}
\mathcal{A} \succ \mathcal{B}_\beta & \xrightarrow{\sigma_{\beta\gamma}} & \mathcal{B}_\gamma \prec \mathcal{A} \\
\uparrow \pi_\beta & & \uparrow \pi_\gamma \\
\bar{\mathcal{B}}_\beta & \xrightarrow{\pi_\gamma^{-1} \circ \sigma_{\beta\gamma} \circ \pi_\beta = \tilde{\sigma}_{\beta\gamma}} & \bar{\mathcal{B}}_\gamma
\end{array}$$

. For every $i \in \omega$ we have

$$\bar{\alpha} \in \pi_\beta(F_i) \iff \bar{\alpha} = \sigma_{\beta\gamma}(\bar{\alpha}) \in \sigma_{\beta\gamma} \circ \pi_\beta(F_i)$$
$$\iff \bar{\alpha} \in \tilde{\sigma}_{\beta\gamma} \circ \pi_\gamma(F_i)$$
$$\iff \tilde{\alpha} \stackrel{\text{def}}{=} \sigma_{\beta\gamma}^{-1}(\bar{\alpha}) \in \pi_\gamma(F_i)$$
$$\iff \tilde{\alpha} \in \{\mu \in \pi_\gamma(\bar{\alpha})\,;\, \pi_\gamma(f_{i+1})(\mu) \in \pi_\gamma(f_i)(\mu)\}.$$

Thus $\pi_\gamma(f_0)(\tilde{\alpha}) \ni \pi_\gamma(f_1)(\tilde{\alpha}) \ni \pi_\gamma(f_2)(\tilde{\alpha}) \ni \ldots$, contradiction.

The canonical embedding from K^{DJ} to $({}^\alpha K^{\mathsf{DJ}} \cap K^{\mathsf{DJ}})/U$ yields an elementary map $\tilde{\pi}$: $K^{\mathsf{DJ}} \to K^{\mathsf{DJ}}$ with critical point $\bar{\alpha}$ (see Lemma 3.34). By Lemma 3.35, there is an inner model with a measurable cardinal $< \tau$. With the same arguments as in Case 1 of the proof of Theorem 3.57 we get that $\kappa \to (<\tau)_2^{<\omega}$ in K^{DJ}.

Case 2. *There is an iterable premouse M of cardinality less than τ, such that for every $\beta < \tau$, $\mathsf{lp}(M) \not\subseteq \bar{\mathcal{B}}_\beta$.*

Let $\langle M_i, \pi_{ij}, \gamma_i, U_i \rangle_{i \leq j < \tau}$ be the τ-iteration of M. For every $\beta < \tau$ let $\bar{\mathcal{B}}_\beta = \langle \bar{B}_\beta, \in, \bar{D}_\beta, g_\beta \rangle$. For $i < \tau$ define $\bar{g}_i \stackrel{\text{def}}{=} g_{\gamma_i} \upharpoonright [\gamma_i]^{<\omega}$.

(6) \hspace{3cm} For every $i < \tau$, $\bar{g}_i \in M_i$.

To see this let $i < \tau$ be arbitrary. Since $\mathcal{A} \models K^{\mathsf{DJ}} = V$, $\bar{\mathcal{B}}_{\gamma_i} \models K^{\mathsf{DJ}} = V$. So there is $N \in \bar{\mathcal{B}}_{\gamma_i}$ such that $\bar{g}_i \in \mathsf{lp}(N)$ and N is an iterable premouse. By Lemma 3.40 it suffices to show that $N_\tau \in M_\tau$. But if $M \subset N_\tau$ then $\mathsf{lp}(M) \subseteq N \subseteq \bar{\mathcal{B}}_{\gamma_i}$, contradiction.

By Lemma 3.41(1), for each $j < \tau$ there are some $x_j \in M$ and $\vec{\rho}_j \in {}^{<\omega}\{\gamma_i\,;\, i < \tau\}$ such that

$$\bar{g}_j = \pi_{0j}(x_j)(\vec{\rho}_j).$$

By Fodor's Lemma there is some stationary set $E \subseteq \tau$ such that for all $j \in E$, $\langle x_j, \vec{\rho}_j\rangle$ is constant, say $\langle x, \vec{\rho}\rangle$. Let

$$C_j \stackrel{\text{def}}{=} \{\gamma_i\,;\, i < j, \gamma_i > \mathsf{max}(\vec{\rho})\}.$$

By Lemma 3.41(3), there is a sequence $\langle k_n\,;\, n < \omega\rangle \in {}^{\omega}2$ such that for every $j \in E$ and every $n < \omega$, $\bar{g}_j``[C_j]^n = \{k_n\}$. Then for every $j < \tau$ the sequence

$$Y_j \stackrel{\text{def}}{=} \pi_{\gamma_j}``C_j$$

is a homogeneous sequence for g in $\mathsf{HOD}[X]$, and each Y_j is a good set of indiscernibles for $\mathcal{A}$ of ordertype j. By taking a subsequence one can ensure that we have a sequence $\langle I_\zeta\,;\, \zeta < \tau\rangle \in \mathsf{HOD}[X]$ such that $\mathsf{ot}(I_\zeta) > \omega(1+\zeta)$, I_ζ is a good set of indiscernibles for $\mathcal{A}$ and the I_ζ agree in the formulas of $\mathcal{A}$ with parameters below τ.

Now, if $\tau = \omega_1$ we continue exactly as in the end of the proof of [**DK83**, Theorem D]. If $\tau > \omega_1$, it means that for every $\zeta < \tau$ there is a $\zeta \leq \zeta' < \tau$ such that $\mathsf{cf}(\mathsf{ot}(I_{\zeta'})) > \omega_1$. Therefore in this case we can directly use Jensen's indiscernibility lemma (Lemma 3.42) and get for every $\zeta < \tau$ a set of indiscernibles in K^{DJ}. By using AC in K^{DJ} we get the desired homogeneous sequence for g of order τ in K^{DJ}. qed

For the infinitary version, note that if $\bigcup_{n\in\omega}\kappa_n = \bigcup_{n\in\omega}\lambda_n$ then

$$(\kappa_n)_{n<\omega} \twoheadrightarrow (\lambda_n)_{n<\omega}$$

implies that $\kappa \stackrel{\text{def}}{=} \bigcup_{n\in\omega}\kappa_n$ is a singular Jónsson cardinal. In [**AK06**, Theorem 6] it is proved that if κ is a singular Jónsson cardinal in a model of ZF then κ is measurable in some inner model. As a corollary to that we get the following.

THEOREM 3.62. *If $\langle\kappa_n\,;\, n \in \omega\rangle$ and $\langle\lambda_n\,;\, n \in \omega\rangle$ are increasing sequences of cardinals such that $\bigcup_{n\in\omega}\kappa_n = \bigcup_{n\in\omega}\lambda_n$, then the infinitary Chang conjecture $(\kappa_n)_{n<\omega} \twoheadrightarrow (\lambda_n)_{n<\omega}$ implies that there is an inner model in which $\kappa = \bigcup_{n\in\omega}\kappa_n$ is measurable.*

Since we can force such a coherent sequence of Erdős cardinals by starting with a measurable cardinal (see [**AK06**, Theorem 3]) we have the following.

COROLLARY 3.63. *The theory* ZF+ *"an infinitary Chang conjecture holds with the supremum of the left hand side cardinals being the same as the supremum of the right hand side cardinals" is equiconsistent with the theory* ZFC+ *"a measurable cardinal exists".*

We conjecture that if the supremum of the κ_n is strictly bigger than the supremum of the λ_n then the consistency strength of such an infinitary Chang conjecture in ZF is weaker. To prove this one would have to look into the details of core model arguments for core models with stronger large cardinal axioms, something that is not in the scope of this thesis.

4. Successor of a singular of cofinality ω

In this case we didn't manage to get equiconsistencies. The reason for this will be discussed in the end of the section. We will show that the consistency strength of ZF + "$(\eta^+, \theta) \twoheadrightarrow (\lambda, \rho)$" + "$\eta$ is singular" is between ZFC + "there exists η a strongly compact cardinal" + "for some $\kappa > \eta$, $\kappa \to^\theta (\lambda)_2^{<\omega}$" and the theory ZFC + "$0^\dagger$ exists". As we mentioned before, this means two things. First that if we have a model of ZFC + "there exists η a strongly compact cardinal" + "for some $\kappa > \eta$, $\kappa \to^\theta (\lambda)_2^{<\omega}$" then we can build a symmetric model in which $(\eta^+, \theta) \twoheadrightarrow (\lambda, \rho)$ and η is singular. Secondly that if we have a model of ZF + "$(\eta^+, \theta) \twoheadrightarrow (\lambda, \rho)$" + "$\eta$ is singular" then there is some inner model of ZFC in which $0^\dagger$ exists.

In this section, the strength required for, and extracted from these principles, comes purely from having a large cardinal the successor of a singular cardinal. So instead of concentrating in the particular large cardinal axioms, we will just look at the example case of the four cardinal Chang conjecture.

To make a cardinal with a partition property be the successor of some given singular we will use the set sized Gitik model with one strongly compact. We will consider two cases. In the first and simpler we will just construct the Gitik model between one strongly compact cardinal and an appropriate partition cardinal. There we will have that, e.g., the Chang conjecture $(\eta^+, \theta) \twoheadrightarrow (\lambda, \rho)$ holds for some singular cardinal η. In the second case we will see how to make the singular η a particular singular we want. We will work with $\aleph_\omega$ but it will not be hard to see how to adapt this construction to work with any other desired singular of cofinality ω. For this second case we will combine the construction of the first case with some cardinal collapses.

To get strength from a Chang conjecture $(\eta^+, \theta) \twoheadrightarrow (\lambda, \rho)$ when η is singular, we will use an idea from [**AK08**].

4.1. Forcing side. This model is basically the model in Section 2 of Chapter 2, for just one strongly compact cardinal. Let $\kappa, \eta, \theta, \lambda$ be infinite cardinals with $\kappa > \eta, \theta, \lambda$. Let $\kappa \to^\theta (\lambda)_2^{<\omega}$ and let η be a $<\kappa$-strongly compact cardinal, i.e., a cardinal such that if $\alpha < \kappa$ then η is α-strongly compact. We want a model where the η is singular and $\kappa = \eta^+$. Remember that $\mathsf{Reg}^{[\eta,\kappa)}$ is the set of all regular cardinals in the interval $[\eta, \kappa)$. For every $\alpha \in \mathsf{Reg}^{[\eta,\kappa)}$, let H_λ be an η-complete, fine ultrafilter over $\mathcal{P}_\eta\alpha$ and let $h_\alpha : \mathcal{P}_\eta\alpha \to \alpha$ be a surjection. Define

$$\Phi_\alpha \stackrel{\text{def}}{=} \{X \subseteq \alpha \,;\, h_\alpha^{-1}``X \in H_\alpha\},$$

where $h_\alpha^{-1}``X = \bigcup\{h_\alpha^{-1}(x) \,;\, x \in X\}$ (since h_α is a surjection, the set $h_\alpha^{-1}(x)$ is a subset of $\mathcal{P}_\eta\alpha$). For every $\alpha \in \mathsf{Reg}^{[\eta,\kappa)}$ the set Φ_α is an η-complete ultrafilter over α. As in Section 2 of Chapter 2, we have defined these ultrafilters in this manner so that we can ensure later that the interval (η, κ) has collapsed to η. Let $\mathbb{P}_\alpha$ be the tree-Prikry forcing for α using Φ_α as defined in Definition 2.1. We force with the partial order $\mathbb{P}$ defined as the finite support product of all $\mathbb{P}_\alpha$ for $\alpha \in \mathsf{Reg}^{[\eta,\kappa)}$.

Intuitively, we will approximate our symmetric model by taking finitely many of these $\mathbb{P}_\alpha$'s.

Formally, for each $\alpha \in \mathsf{Reg}^{[\eta,\kappa)}$ let $\mathcal{G}_\alpha$ be the group of permutations of α that move only finitely many elements of α. Let $\mathcal{G}$ be the finite support product of all these $\mathcal{G}_\alpha$'s. For $\vec{T} \in \mathbb{P}$ and $a = \langle a_\alpha \,;\, \alpha \in \mathsf{supp}a\rangle \in \mathcal{G}$ define

$$a\vec{T} \stackrel{\text{def}}{=} \langle \{a_\alpha``t \,;\, t \in T_\alpha\} \,;\, \alpha \in \mathsf{dom}\vec{T}\rangle.$$

This map a is an automorphism of $\mathbb{P}$ and we view $\mathcal{G}$ as an automorphism group of $\mathbb{P}$. If e is a finite subset of $\mathsf{Reg}^{[\eta,\kappa)}$ then let $E_e \stackrel{\text{def}}{=} \{\vec{P}\restriction e \,;\, \vec{P} \in \mathbb{P}\}$. The set $I \stackrel{\text{def}}{=} \{E_e \,;$ e is a finite subset of $\mathsf{Reg}^{[\eta,\kappa)}\}$ is a symmetry generator with $\vec{P}\restriction^* E_e = \vec{P}\restriction e$.

Let $V(G) = V(G)^{\mathcal{G},\mathcal{F}_I}$. As we said this model is just a special case of the model presented in Section 2 of Chapter 2. We wrote the definition again so that it is easier to use in the next subsection. As a special case of that model, it has the same properties.

So the approximation lemma holds for $V(G)$, in which η is a singular cardinal, the cardinals below η are the same as the cardinals below η in the ground model, and there are no cardinals in the interval (η, κ). So it remains to show the following lemma.

PROPOSITION 3.64. *In $V(G)$, κ is a cardinal and $\kappa \to^\theta (\lambda)^{<\omega}_2$.*

PROOF. Assume for a contradiction that it isn't. Then for some $\beta < \kappa$ there is some injection $f : \kappa \to \beta$. Let $\dot{f} \in \mathsf{HS}$ be a name for f with support $e \subset \mathsf{Reg}^{[\eta,\kappa)}$. Then $f \in V[G \cap E_e]$ which is impossible since by inaccessibility of κ, E_e has cardinality less than κ and thus the κ-cc.

Lemma 3.45 says that the property $\kappa \to^\theta (\lambda)^{<\omega}_2$ is preserved under forcing with a partial order of cardinality less than κ. So with a similar argument to the one above we have that in $V(G)$, $\kappa \to^\theta (\lambda)^{<\omega}_2$.

qed

By Lemma 3.4 we get as a corollary that in $V(G)$, for any $\theta < \kappa$ and $\rho < \lambda$, $(\eta^+, \theta) \twoheadrightarrow (\lambda, \rho)$ and η is singular.

4.2. Chang conjectures starting with $\aleph_\omega$. In order to have a Chang conjecture for the successor of a particular singular cardinal, we do the construction of the previous subsection while simultaneously collapsing the Prikry sequence that is being created to become $\omega, \omega_2, \omega_4, \ldots$ etc..

So let $\kappa, \eta, \theta, \lambda$ be infinite cardinals such that $\kappa > \eta, \theta, \lambda$. Assume $\kappa = \kappa(\lambda)$ and η is $< \kappa$-strongly compact, i.e., for every $\alpha \in [\eta, \kappa)$, η is α-strongly compact.

Let $\mathsf{in}^{[\eta,\kappa)}$ be the set of all inaccessible cardinals in $[\eta, \kappa)$. For every $\alpha \in \mathsf{in}^{[\eta,\kappa)}$, let U_α be a fine ultrafilter over $\mathcal{P}_\eta(\alpha)$, and let $h_\alpha : \mathcal{P}_\eta(\alpha) \to \alpha$ be a bijection. For every $\alpha \in \mathsf{in}^{[\eta,\kappa)}$ define

$$\Phi_\alpha \stackrel{\text{def}}{=} \{x \subseteq \alpha \,;\, h_\alpha``x \in H_\alpha\},$$

which is an η-complete ultrafilter over α. Let $\mathbb{P}_\alpha$ be the injective tree-Prikry forcing with respect to Φ_α, as defined in Definition 2.1. Let

$$\mathbb{P} \stackrel{\text{def}}{=} \mathbb{P}_\eta \times \prod^{\mathsf{fin}}_{\alpha \in (\mathsf{in}^{[\eta,\kappa)} \setminus \eta)} \mathbb{P}_\alpha.$$

We will define a $\mathbb{P}$-name $\langle \dot{\mathbb{Q}}, \leq_{\dot{\mathbb{Q}}}, 1_{\dot{\mathbb{Q}}} \rangle$ for a partial order. To define the set, first note that we will use "$\mathsf{tuple}(\check{x}_1, \ldots, \check{x}_n)$" for the canonical name for a tuple of names.

$$\begin{aligned} \left\langle \sigma, \vec{T} \right\rangle \in \dot{\mathbb{Q}} \stackrel{\text{def}}{\Longleftrightarrow} & \vec{T} \in \mathbb{P} \text{ and} \\ & \exists n \in \omega (\sigma = \mathsf{tuple}(\check{q}_0, \check{q}_1, \ldots, \check{q}_n) \text{ and } n = \mathsf{dom}(T) \text{ and} \\ & (\text{if } 0 \in \mathsf{dom}(T) \text{ then } q_0 \in \mathsf{Fn}(\omega, T(0), \omega)) \text{ and} \\ & (\text{if } 0 < i \in \mathsf{dom}(T) \text{ then } q_i \in \mathsf{Fn}(T(i-1)^+, T(i), T(i-1)^+))) \end{aligned}$$

To define the partial order on $\dot{\mathbb{Q}}$, for $\left\langle \mathsf{tuple}(\check{q}_0, \ldots, \check{q}_n), \vec{T} \right\rangle$, $\left\langle \mathsf{tuple}(\check{r}_0, \ldots, \check{r}_m), \vec{S} \right\rangle \in \dot{\mathbb{Q}}$, define

$$\vec{T} \Vdash \mathsf{tuple}(\check{q}_0, \ldots, \check{q}_n) \leq_{\dot{\mathbb{Q}}} \mathsf{tuple}(\check{r}_0, \ldots, \check{r}_m) \Longleftrightarrow n \geq m \text{ and } \forall i = 0, \ldots, m (q_i \supseteq r_i).$$

Let $1_{\dot{\mathbb{Q}}} \stackrel{\text{def}}{=} \varnothing$. We will force with the iteration

$$\mathbb{P} * \dot{\mathbb{Q}}.$$

Now define $i : \mathbb{P} \to \mathbb{P} * \dot{\mathbb{Q}}$ by $i(\vec{T}) = \left\langle \vec{T}, 1_{\dot{\mathbb{Q}}} \right\rangle$. Let $G_{\mathbb{P}} \stackrel{\text{def}}{=} i^{-1}(G)$ and let

$$G_{\dot{\mathbb{Q}}} \stackrel{\text{def}}{=} \{ \tau^{G_{\mathbb{P}}} \; ; \; \tau \in \mathsf{dom}(\dot{\mathbb{Q}}) \wedge \exists \vec{T} (\left\langle \vec{T}, \tau \right\rangle \in G) \}.$$

By [**Kun80**, Theorem 5.5], $G_{\mathbb{P}}$ is $\mathbb{P}$-generic, $G_{\dot{\mathbb{Q}}}$ is $\dot{\mathbb{Q}}^{G_{\mathbb{P}}}$- generic over $V[G_{\mathbb{P}}]$, $G = G_{\mathbb{P}} * G_{\dot{\mathbb{Q}}}$, and $V[G] = V[G_{\mathbb{P}}][G_{\dot{\mathbb{Q}}}]$. In $V[G_{\mathbb{P}}]$, η is singular and nothing has collapsed $\leq \eta$. Say the Prikry sequence that has been added is $\langle \gamma_0, \gamma_1, \ldots \rangle$. It's easy to check that

$$\dot{\mathbb{Q}}^{G_{\mathbb{P}}} = \mathsf{Fn}(\omega, \gamma_0, \omega) \times \prod_{0<i<\omega} \mathsf{Fn}(\gamma_{i-1}^+, \gamma_i, \gamma_{i-1}^+).$$

For each $\alpha \in (\mathsf{in}^{[\eta,\kappa)} \setminus \eta)$ let $\mathcal{G}_\alpha$ be the group of permutations of α that only move finitely many elements of α. Let $\mathcal{G}$ be the finite support product of all $\mathcal{G}_\alpha$, i.e.,

$$\mathcal{G} = \prod_{\alpha \in (\mathsf{in}^{[\eta,\kappa)} \setminus \eta)}^{\mathsf{fin}} \mathcal{G}_\alpha.$$

For convenience we denote an $a \in \mathcal{G}$ by $a = \langle a_\alpha \; ; \; \alpha \in (\mathsf{in}^{[\eta,\kappa)} \setminus \eta) \rangle$, with $a_\alpha = \mathsf{id}$ for all but finitely many α. For $(\vec{T}, \sigma) \in \mathbb{P} \star \dot{\mathbb{Q}}$, and $a \in \mathcal{G}$, if $\vec{T} = \langle T_0 \rangle^\frown \langle T_e \; ; \; e \subseteq_{\mathsf{fin}} (\mathsf{in}^{[\eta,\kappa)} \setminus \eta) \rangle$, then define

$$a(\left\langle \vec{T}, \sigma \right\rangle) \stackrel{\text{def}}{=} \langle T_0 \rangle^\frown \langle a_\alpha {}^{\text{``}}T_\alpha \; ; \; \alpha \in \mathsf{dom}(\vec{T}) \rangle^\frown \langle \sigma \rangle \, ,$$

where $a_\alpha {}^{\text{``}}T_\alpha = \{ \{ (n, a_\alpha(\beta)) \; ; \; (n, \beta) \in t \} t \in T_\alpha \; ; \; \}$ as defined in Definition 2.8. This map is now an automorphism group of $\mathbb{P} * \dot{\mathbb{Q}}$ for the same reasons as for Proposition 2.9.

For each finite $e \subseteq (\mathsf{in}^{[\eta,\kappa)} \setminus \eta)$ define

$$\begin{aligned} E_e & \stackrel{\text{def}}{=} \{ \left\langle \vec{T}, \sigma \right\rangle \in \mathbb{P} * \dot{\mathbb{Q}} \; ; \; \mathsf{dom}(\vec{T}) = \{\eta\} \cup e \}, \text{ and} \\ I & \stackrel{\text{def}}{=} \{ E_e \; ; \; e \subseteq (\mathsf{in}^{[\eta,\kappa)} \setminus \eta) \text{ is finite} \}. \end{aligned}$$

This is a projectable symmetry generator with projections

$$\left\langle \vec{T}, \sigma \right\rangle \restriction^* E_e = \langle T_\eta \rangle^\frown \langle T_\alpha \; ; \; \alpha \in e \rangle^\frown \langle \sigma \rangle \, .$$

Take the symmetric model $V(G) \stackrel{\text{def}}{=} V(G)^{\mathcal{F}_I}$. It's not hard to check that the approximation lemma holds for $V(G)$, i.e., $\mathbb{P} * \dot{\mathbb{Q}}$ is $\mathcal{G}, I$-homogeneous.

LEMMA 1. *In $V(G)$, $\eta = \aleph_\omega$.*

PROOF. In $V[G \restriction E_\varnothing]$ everything below η is below $\aleph_\omega$, and every $E_\varnothing$-name is a name in HS. So it suffices to show that η has not collapsed. Assume towards contradiction that for some $\beta < \eta$ there is a bijection $f : \beta \to \eta$. Let $\dot{f} \in \mathsf{HS}$ be a $\mathbb{P} * \dot{\mathbb{Q}}$-name for f with support $E_e \in I$. By the approximation lemma for $V(G)$, $f \in V[G \cap E_e]$. Let $e = \{\alpha_1, \ldots, \alpha_m\}$ and note that

$$E_e = (\mathbb{P}_\eta \times \mathbb{P}_{\alpha_1} \times \cdots \times \mathbb{P}_{\alpha_m}) * \dot{\mathbb{Q}} \stackrel{\text{def}}{=} \mathbb{P}' * \dot{\mathbb{Q}}.$$

With a Prikry lemma just as Lemma 2.18, we can show that $\langle \mathbb{P}', \leq^*_{\mathbb{P}'} \rangle$ is η-closed so just like in Theorem 2.19 we can show that $\mathbb{P}'$ preserves η. So f must be added by $\dot{\mathbb{Q}}$. But for any $\mathbb{P}'$-generic filter G',

$$\dot{\mathbb{Q}}^{G'} = \mathsf{Fn}(\omega, \gamma_0, \omega) \times \prod_{0<i<\omega} \mathsf{Fn}(\gamma_{i-1}^+, \gamma_i, \gamma_{i-1}^+),$$

and this preserves every γ_{i-1}^+, and the sequence $\langle \gamma_{i-1}^+ \ ; \ i \in \omega \rangle$ is cofinal to η. Contradiction. qed

Having this lemma and by a small forcing argument we get the following.

COROLLARY 3.65. *In $V(G)[H]$, $\eta = \aleph_\omega$ and $\eta^+ \to^\theta (\lambda)_2^{<\omega}$.*

Therefore we have constructed a model of

$$(\aleph_{\omega+1}, |\theta|) \twoheadrightarrow (|\lambda|, |\rho|).$$

As before, we could use this construction to replace the partition property with measurability, weak compactness, Ramseyness, etc. This again is based on that the construction involves only small symmetric forcing with respect to the large cardinal in mention.

4.3. Getting strength from successors of singular cardinals. In this section we will start from the premise of ZF + "$(\eta^+, \theta) \twoheadrightarrow (\lambda, \rho)$" + "$\eta$ is singular". We will distinguish two cases. First if $(\eta^+)^{\mathsf{HOD}} < \eta^+$, then we will see that there is an inner model with a class of strong cardinals.

DEFINITION 3.66. A cardinal μ is a strong cardinal if for every γ, there is a non-trivial elementary embedding $j : V \to M$ with critical point μ and such that $\gamma < j(\mu)$, and $V_{\mu+\gamma} \subseteq M$.

This appears to be a statement about classes. But these elementary embeddings can be approximated by certain sets called *extenders*. It's not our purpose to discuss these issues so we refer the reader to [**Kan03**, §26 "Extenders"]. In terms of consistency strength, a strong cardinal is weaker than a supercompact [**Kan03**, Theorem 26.14] and stronger than a measurable [**Kan03**, Exercise 26.6].

To prove our result we use the core model below a class of strong cardinals, developed by Ralf Schindler and presented in [**Sch02**]. We will denote this core model by K. According to [**AK08**, Proposition 11], here we also have that for every set $a \subseteq \mathsf{HOD}$, $K^{\mathsf{HOD}} = K^{\mathsf{HOD}[a]}$.

According to [**Sch02**, Theorem 8.18], if there is no inner model with a class of strong cardinals and K is built in a model V of ZFC then in V the following covering property holds:

$$\text{if } \kappa \geq \omega_2, \text{ then } \mathsf{cf}((\kappa^+)^K) \geq |\kappa| \tag{7}$$

Using this property above we can prove our result in quite exactly as is done in [**AK08**, Lemma 4].

LEMMA 3.67. *Let V be a model of ZF. If $\eta \leq \theta, \lambda < \rho$ are infinite cardinals, η is singular, $(\eta^+)^{\mathsf{HOD}} < \eta^+$, and $(\eta^+, \theta) \twoheadrightarrow (\lambda, \rho)$, then there is an inner model with a class of strong cardinals.*

Now, if $(\eta^+)^{\mathsf{HOD}} = \eta^+$ then we will look at the literature, in the proofs which assume the axiom of choice. These core model proofs will involve getting a contradiction from the existence of a certain structure. In many cases we can work with $\mathsf{HOD}[x]$, where x is a set of ordinals that witnesses the structure that will give the contradiction. Let's see an example. Levinski's proof of [**Lev84**, Theorem B] shows that for any singular cardinal κ, the Chang conjecture $(\kappa^+, \kappa) \twoheadrightarrow (\omega_1, \omega)$ implies $0^\dagger$. Let's see as an example how to modify this proof for our choiceless case.

LEMMA 3.68. *Assume that there is a model of ZF+ "there are cardinals η, λ, ρ such that $(\eta^+, \eta) \twoheadrightarrow (\lambda, \rho)$, η is singular, and $(\eta^+)^{\mathsf{HOD}} = \eta^+$" Then $0^\dagger$ exists.*

PROOF. Let $\kappa \stackrel{\text{def}}{=} (\eta^+)^V = (\eta^+)^{\mathsf{HOD}}$ and consider the structure $\langle K_\kappa, \eta, \in\rangle$. Use our Chang conjecture to get an elementary substructure

$$\langle K', K' \cap \theta, \in\rangle \prec \langle K_\kappa, \eta, \in\rangle,$$

such that $|K'| = \lambda$ and $|K' \cap \theta| = \rho$. Let X be a cofinal subset of η, $|X| = \mathsf{cf}\eta$, and let x be a set of ordinals coding $X \times K'$. We attach x to HOD getting $\mathsf{HOD}[x]$. By [**AK06**, Proposition 1.1], $K^{\mathsf{HOD}} = K^{\mathsf{HOD}[x]}$. We are now working in $\mathsf{HOD}[x]$.

Assume towards a contradiction that K_κ covers $(V_\kappa)^{\mathsf{HOD}[x]}$. By [**DK83**, Lemma 2.4] η is still singular in K_κ and $(\eta^+)^K = \kappa$.

We want to show that there exists an inner model with a measurable cardinal below κ. Let $\langle \bar{K}, A, \in\rangle$ be the Mostowski collapse of $\langle K', K' \cap \theta, \in\rangle$, with $\pi : \bar{K} \to K'$ the isomorphism. Let

$$\alpha \stackrel{\text{def}}{=} \pi^{-1}(\eta).$$

Note that $\rho = \mathsf{ot}(\bar{K} \cap \eta) = \pi^{-1}(\bar{K} \cap \eta)$, and $|\bar{K}| = \lambda$.

It's not hard to see that π is a non-trivial elementary embedding whose critical point, say $\gamma \stackrel{\text{def}}{=} \mathsf{crit}\pi$ is less than or equal to α.

Now, since η is the largest cardinal in $\langle K_\kappa, \eta, \in\rangle$ we have that $\bar{K} \models$"α is the largest cardinal". Thus $\mathrm{Ord} \cap \bar{K} = \alpha^+$ and by [**Lev84**, Lemma 17] we get $\bar{K} = K_{\alpha^+}$. By [**Lev84**, Proposition 14] we get that there is an inner model with a measurable cardinal μ such that

(1) if $\omega_1 \leq \alpha$ then $\mu < \gamma^+$, and
(2) if $\gamma < \omega_1$ then $\mu \leq \omega_1$.

In both cases we have that $\mu < \eta^+$, which is a contradiction. Thus $0^\dagger$ exists. qed

This proof was presented here because in the literature it is the strongest proof that involves the successor of a singular cardinal in a Chang conjecture. The state of core model theory can certainly give more strength from such a Chang conjecture but it is outside of the purposes of this thesis to get into the details of higher core models.

4

Conclusions, open questions and future research

In Chapter 1 we developed a "mechanised" technique for creating symmetric models. We expanded on what is already known about symmetric forcing by isolating useful properties that symmetric models can have, such as satisfying the approximation lemma. All the symmetric models that we constructed satisfy the approximation lemma. In that procedure we were accompanied by an example, the Feferman-Lévy model in which the set of reals is a countable union of countable sets.

Then we saw how to make successor cardinals with large cardinal properties in models of ZF. We once more "mechanised" Jech's construction of [**Jec68**] so that we can use it in a "passe-partout" fashion in the rest of the chapters. With this construction we can only get measurable cardinals with normal measures. That left the question of whether there exists a model of ZF+"there exists a measurable cardinal without a normal measure"? During the end phase of the thesis the author received a preprint of a paper from Moti Gitik with Eilon Bilinsky in which they indeed produce such a model.

In the end of that chapter we demonstrated how this passe-partout form can be used to do multiple collapses simultaneously. For that we constructed, for some ρ, a model of ZF+"there exists a ρ-long sequence such that every second element of it is measurable"$+\neg\mathsf{AC}_{\kappa_1}(\mathcal{P}(\kappa_0))$. In that model we could not decide the following question.

QUESTION 1.
In the first model in Section 4 of Chapter 1, does DC *fail?*

We mentioned in the end how to make sure that DC holds for another symmetric model by using a different symmetry generator.

In Chapter 2 we turned our attention to singular cardinals and several patterns of them. Since Jech's method from Chapter 1 cannot be used to collapse on a singular cardinal, we started by presenting a modification of tree-Prikry forcing and of strongly compact tree Prikry forcing. These modifications are very similar to the original versions but the next question stays undecided.

QUESTION 2.
Does injective (strongly compact) tree-Prikry forcing completely or even densely embed into (strongly compact) tree-Prikry forcing, or the other way around?

We started our investigation of patterns of singular cardinals with the simplest non trivial pattern; starting from a model of ZFC we got an arbitrarily long sequence of alternating regular and singular cardinals. This we could still do with just the construction of Jech, since there are no successive singulars involved.

The next section was the first place we saw (potential) successive singulars. There we started from a model of ZFC+"there exists a countable sequence of strongly compact cardinals", with a function $f : \omega \to 2$. We constructed a model in which ω_{n+1} is regular iff $f(n) = 1$. The method we used was a combination of symmetric injective tree-Prikry forcing and the symmetric collapses of the Jech model. This is done for just a countable sequence but this method can be implemented in the following constructions of that chapter to make arbitrary patterns of regular and singular cardinals, at any length.

We saw then how to get over limits of such sequences and constructed a ρ-long sequence of successive singular cardinals, where ρ is any ordinal less than or equal to ω_1 of the ground model. We proved that in that symmetric model there is a ρ-long sequence of successive singular cardinals, which are all almost Ramsey. We showed that all limit cardinals in that sequence are Rowbottom cardinals carrying Rowbottom filters and we asked the following question.

QUESTION 3.
Could one modify the construction in Section 4 of Chapter 2, without adding consistency strength to the assumptions, in order to get a symmetric model in which all cardinals in (ω, η) are Rowbottom cardinals carrying Rowbottom filters?

The last section of that chapter was a modification of Gitik's construction in his paper "All uncountable cardinals can be singular". Our version of the construction starts from a model of ZFC plus a ρ-long sequence of strongly compact cardinals without regular limits, for any ρ. We also assumed that there is a measurable cardinal κ above all the strongly compact cardinals. We used specially defined ultrafilters in the definition of our forcing to ensure that the former strongly compact cardinals and their singular limits are the only cardinals left in the interval (ω, κ), in the symmetric model $V(G)$. The measurable cardinal κ remains measurable in the $V(G)$ so it is $\aleph_{\rho+1}$, measurable, and the first regular cardinal as well.

In the beginning of the section already we mentioned that if the measurable were the limit of the entire sequence of strongly compact cardinals, and we skipped the last part of the forcing, then we would have a regular limit cardinal being the first regular cardinal and the first measurable cardinal in $V(G)$. As it was the case in the previous two sections, all the cardinals in $(\omega, \kappa]$ are almost Ramsey and the singular limit cardinals in that interval are Rowbottom cardinals carrying Rowbottom filters.

In all the models of this chapter, the singulars that are 'created' have all cofinality ω. This is due to the nature of the forcing we are using, which is basically Prikry forcing for adding a countable cofinal sequence to a measurable cardinal. Prikry forcing works here because it doesn't add bounded subsets to the cardinal it is applied to, and that is crucial in showing that the strongly compact cardinals do not collapse in the symmetric model. It is a future project to construct sequences of successive singular cardinals of larger cofinality with Radin forcing, because Radin forcing also does not add bounded subsets to the cardinal it is applied to.

Finally in Chapter 4, we turned to lower consistency strengths, in the realm of Erdős cardinals. Inspired by the proof of the equiconsistency of the existence of an Erdős cardinal with the original Chang conjecture $(\omega_2, \omega_1) \twoheadrightarrow (\omega_1, \omega)$, we proved that for any regular cardinal η, a Chang conjecture of the form $(\eta^+, \theta) \twoheadrightarrow (\lambda, \rho)$ plus ZF is equiconsistent with the existence of the Erdős cardinal $\kappa(\lambda)$ in a model of ZFC.

We looked at longer Chang conjectures of the form $(\kappa_n, \dots, \kappa_1) \twoheadrightarrow (\lambda_n, \dots, \lambda_1)$ and proved that such a Chang conjecture plus ZF is equiconsistent with the existence of $(n-1)$-many Erdős cardinals in a model of ZFC.

The longest Chang conjecture considered in the literature is the infinitary Chang conjecture $(\kappa_n)_{n<\omega} \twoheadrightarrow (\lambda_n)_{n<\omega}$ for which we proved that if κ is the limit of both the κ_n and λ_n then this infinitary Chang conjecture plus ZF is equiconsistent with the existence of a measurable cardinal in a model of ZFC. The following question remains open.

QUESTION 4.
What is the consistency strength of the theory ZF+ *the infinitary Chang conjecture* $(\kappa_n)_{n\in\omega} \twoheadrightarrow (\lambda_n)_{n\in\omega}$*, when* $\bigcup_{0<n<\omega} \kappa_n > \bigcup_{0<n<\omega} \lambda_n$*?*

Lastly we involved a successor of a singular cardinal η of cofinality ω in a Chang conjecture of the form $(\eta^+, \theta) \twoheadrightarrow (\lambda, \rho)$. We showed that if we start from a model of ZFC plus a strongly compact cardinal η with a λ-Erdős cardinal on top, then we can construct a model of ZF in which $(\eta^+, \theta) \twoheadrightarrow (\lambda, \rho)$ holds, for any $\theta \leq \eta$ and $\rho < \lambda$. We then constructed a model in which the above holds and η is $\aleph_\omega$, with a sort of symmetric forcing iteration of injective tree-Prikry and a collapse of the Prikry sequence to the ω_n.

On the other side we got a lower bound as high as the present techniques in core model theory can provide. We did not give any detailed proofs in that last subsection, just a sketch on the way one would go about to get such lower bounds. To produce detailed proofs of these lower bounds is another project for the future. At this end point we should mention that we expect that when a core model for a supercompact cardinals is developed, this gap in consistency strength will close.

Index

Bibliography

[ADK] Arthur Apter, Ioanna M. Dimitríou, and Peter Koepke. The first measurable and first regular cardinal can simultaneously be $\aleph_{\rho+1}$, for any ρ. Submitted for publication.

[AH91] Arthur W. Apter and James M. Henle. Relative consistency results via strong compactness. *Fundamenta Mathematicae*, 139:133–149, 1991.

[AJL] Arthur Apter, Stephen Jackson, and Benedikt Löwe. Cofinality and measurability of the first three uncountable cardinals. to appear in Transactions of the American Mathematical Society.

[AK06] Arthur W. Apter and Peter Koepke. The consistency strength of $\aleph_\omega$ and $\aleph_{\omega_1}$ being Rowbottom cardinals without the axiom of choice. *Archive for Mathematical Logic*, 45:721–737, 2006.

[AK08] Arthur Apter and Peter Koepke. Making all cardinals almost Ramsey. *Archive for Mathematical Logic*, 47:769–783, 2008.

[Apt83a] Arthur Apter. Some results on consecutive large cardinals. *Annals of Pure and Applied Logic*, 25:1–17, 1983.

[Apt83b] Arthur W. Apter. On a problem of Silver. *Fundamenta Mathematicae*, 116:33–38, 1983.

[Apt85] Arthur W. Apter. An AD-like model. *Journal of Symbolic Logic*, 50:531–543, 1985.

[Apt96] Arthur W. Apter. AD and patterns of singular cardinals below Θ. *Journal of Symbolic Logic*, 61:225–235, 1996.

[BDL06] Andreas Blass, Ioanna M. Dimitríou, and Benedikt Löwe. Inaccessible cardinals without the axiom of choice. *Fundamenta Mathematicae*, 194:179–189, 2006.

[Bol51] Bernard Bolzano. *Paradoxien des unendlichen.* Felix Meiner, 1851. Digitalized from an original from the University of Michigan, url: `http://books.google.com/books?id=RT84AAAAMAAJ`.

[Bul78] Everett L. Bull, Jr. Consecutive large cardinals. *Annals of Mathematical Logic*, 15:161–191, 1978.

[CK90] Chen Chung Chang and H. Jerome Keisler. *Model theory*, volume 58 of *Studies in Logic and the Foundations of Mathematics*. North Holland, 3rd edition, 1990.

[Coh63] Paul J. Cohen. The independence of the continuum hypothesis. *Proceedings of the National Academy of Sciences of the USA*, 50:1143–1148, 1963.

[Coh64] Paul J. Cohen. The independence of the continuum hypothesis ii. *Proceedings of the National Academy of Sciences of the USA*, 51:105–110, 1964.

[Dim06] Ioanna M. Dimitríou. Strong limits and inaccessibility with non-wellorderable powersets. Master of logic thesis, ILLC, Universiteit van Amsterdam, 2006. Supervised by Dr.Benedikt Löwe (ILLC publication series MoL-2006-3).

[DJ81] Anthony Dodd and Ronald B. Jensen. The core model. *Annals of Mathematical Logic*, 20:43–75, 1981.

[DJK79] Hans D. Donder, Ronald B. Jensen, and Bernd Koppelberg. Some applications of the core model. In Ronald B. Jensen, editor, *Set theory and model theory*, volume 872 of *Lecture Notes in Mathematics*, pages 55–97. Springer, 1979.

[DK83] Hans D. Donder and Peter Koepke. On the consistency strength of "accessible" Jónsson cardinals and of the weak Chang conjecture. *Annals of Pure and Applied Logic*, 25:233–261, 1983.

[FL63] Solovay Feferman and Azriel Lévy. Independence results in set theory by Cohen's method, ii (abstract). *Notices of the American Mathematical Society*, 10:593, 1963.

[For10] Matthew Foreman. Chapter 13: Ideals and generic elementary embeddings. In Matthew Foreman and Akihiro Kanamori, editors, *Handbook of set theory*, pages 885–1147. Springer, 2010.

[Fra22a] A. Abraham H. Fraenkel. Der Begriff 'definit' und die Unabhängigkeit des Auswahlsaxioms. *Sitzungsberichte der Preussischen Akademie der Wissenschaften, Physikalisch-mathematische Klasse*, pages 253–257, 1922.

[Fra22b] A. Abraham H. Fraenkel. Zu den Grundlagen der Cantor-Zermeloschen Mengelehre. *Mathematische Annalen*, 86:230–237, 1922.

[Git80] Moti Gitik. All uncountable cardinals can be singular. *Israel Journal of Mathematics*, 35:61–88, 1980.

[Git10] Moti Gitik. Chapter 16: Prikry-type forcings. In Matthew Foreman and Akihiro Kanamori, editors, *Handbook of set theory*, pages 1351–1447. Springer, 2010.

[Göd30] Kurt Gödel. Die Vollständigkeit der Axiome des logischen Functionenkalküls. *Monatshefte für Mathematik und Physik*, 37:349–360, 1930.

[Göd31] Kurt Gödel. Über formal unentscheidbare Sätze der Principia Mathematica und verwandter Systeme, i. *Monatshefte für Mathematik und Physik*, 38:173–198, 1931.

[Göd38] Kurt Gödel. The consistency of the axiom of choice and of the generalized continuum-hypothesis. *Proceedings of the National Academy of Sciences of the USA*, 24:556–557, 1938.

[Göd39] Kurt Gödel. Consistency proof for the generalized continuum hypothesis. *Proceedings of the National Academy of Sciences of the USA*, 25:220–224, 1939.

[Hau08] Felix Hausdorff. Grundzüge einer Theorie der geordneten Mengen. *Mathematische Annalen*, 65:435–505, 1908.

[Hau14] Felix Hausdorff. *Grundzüge der Mengenlehre*. Veit & Comp., 1914.

[HKR+01] Paul Howard, Kyriakos Keremedis, Jean E. Rubin, Adrienne Stanley, and Eleftherios Tachtsis. Non-constructive properties of the real line. *Mathematical Logic Quarterly*, 47:423–431, 2001.

[Hod97] Wilfrid Hodges. *A shorter model theory*. Cambridge university press, 1997.

[Jec68] Thomas J. Jech. ω_1 can be measurable. *Israel Journal of Mathematics*, 6:363–367, 1968.

[Jec71] Thomas J. Jech. *Lectures in set theory, with particular emphasis on the method of forcing*, volume 217 of *Lecture Notes in Mathematics*. Springer, 1971.

[Jec73] Thomas J. Jech. *The axiom of choice*. Dover publications, INC., 1973.

[Jec78] Thomas J. Jech. *Set theory*. New York Academic Press, 1st edition, 1978.

[Jec02] Thomas Jech. Set theory. *The Stanford Encyclopedia of Philosophy*, 2002. forthcoming URL http://plato.stanford.edu/archives/fall2010/entries/set-theory.

[Jec03] Thomas J. Jech. *Set theory. The third millenium edition, revised and expanded*. Springer, 2003.

[Kan03] Akihiro Kanamori. *The higher infinite*. Springer, 2nd edition, 2003.

[Koe88] Peter Koepke. Some applications of short core models. *Annals of Pure and Applied Logic*, 37(2):179–204, 1988.

[Kun78] Kenneth Kunen. Saturated ideals. *Journal of symbolic logic*, 43:65–76, 1978.

[Kun80] Kenneth Kunen. *Set theory: an introduction to independence proofs*. Elsevier, 1980.

[Lev84] Jean Pierre Levinski. Instances of the conjecture of Chang. *Israel journal of mathematics*, 48(2–3):225–243, 1984.

[LM38] Adolf Lindenbaum and Andrzej Mostowski. Über die Unabhängigkeit des Auswahlsaxioms und einiger seiner Folgerungen. *Comptes Rendus des Séances de la Société des Sciences et des Lettres de Varsovie*, 31:27–32, 1938.

[LMS90] Jean Pierre Levinski, Menachem Magidor, and Saharon Shelah. Chang's conjecture for $\aleph_\omega$. *Israel journal of mathematics*, 69(2):161–172, 1990.

[LS67] Azriel Lévy and Robert M. Solovay. Measurable cardinals and the continuum hypothesis. *Israel Journal of Mathematics*, 5:234–248, 1967.

[Mos39] Andrzej Mostowski. Über die Unabhängigkeit des Wohlordnungssatzes vom Ordnungsprinzip. *Fundamenta Mathematicae*, 32:201–252, 1939.

[Rus19] Bertrand Russel Russell. *The philosophy of logical atomism.* Deptartment of Philosophy, University of Minnesota, 1919.

[Sch99] Ralf D. Schindler. Successive weakly compact or singular cardinals. *Journal of Symbolic Logic*, 64(1):139–146, 1999.

[Sch02] Ralf Schindler. The core model for almost linear iterations. *Annals of pure and applied logic*, 116:205–272, 2002.

[Sco67] Dana S. Scott. A proof of the independence of the continuum hypothesis. *Mathematical Systems Theory*, 1:89–111, 1967.

[She80] Saharon Shelah. A note on cardinal exponentiation. *Journal of Symbolic Logic*, 45:56–66, 1980.

[Sol63] Robert M. Solovay. Independence results in the theory of cardinals (abstract). *Notices of the American Mathematical Society*, 10:595, 1963.

[Spe57] Ernst P. Specker. Zur Axiomatik der Mengenlehre (Fundierungs und Auswahlaxiom). *Zeitschift fur mathematische Logik und Grundlagen der Mathematik*, 3:173–210, 1957.

[Tak70] Gaisi Takeuti. A relativization of strong axioms of infinity to ω_1. *Annals of the Japan Association for Philosophy of Science*, 3:191–204, 1970.

[Vau63] Robert L. Vaught. Models of complete theories. *Bulletin of the American Mathematical Society*, 69:299–313, 1963.

[VH67] Petr Vopěnka and Petr Hájek. Concerning the ∇-models of set theory. *Bulletin de l'Académie Polonaise des Sciences*, 15:113–117, 1967.

[Vop65] Petr Vopěnka. On ∇-model of set theory. *Bulletin de l'Académie Polonaise des Sciences*, 13:267–272, 1965.

[Zer08] Ernst Zermelo. Untersuchungen über die Grundlagen der Mengenlehre. *Mathematische Annalen*, 65(2):261–281, 1908.

MIX
Papier aus verantwortungsvollen Quellen
Paper from responsible sources
FSC® C105338

Printed by Books on Demand GmbH, Norderstedt / Germany